Approaches to Global Sustainability, Markets, and Governance

Series Editors

Approaches to Global Sustainability, Markets, and Governance takes a fresh and global approach to issues of corporate social responsibility, regulation, governance, and sustainability. It encompasses such issues as: environmental sustainability and managing the resources of the world; geopolitics and sustainability; global markets and their regulation; governance and the role of supranational bodies; sustainable production and resource acquisition; society and sustainability.

Although primarily a business and management series, it is interdisciplinary and includes contributions from the social sciences, technology, engineering, politics, philosophy, and other disciplines. It focuses on the issues at a meta-level, and investigates the ideas, organisation, and infrastructure required to address them.

The series is grounded in the belief that any global consideration of sustainability must include such issues as governance, regulation, geopolitics, the environment, and economic activity in combination to recognise the issues and develop solutions for the planet. At present such global meta-analysis is rare as current research assumes that the identification of local best practice will lead to solutions, and individual disciplines act in isolation rather than being combined to identify truly global issues and solutions.

David Crowther · Shahla Seifi
Editors

Preparing for a Sustainable Future

Springer

Editors
David Crowther
Social Responsibility Research Network (SRRNet), SRRNet.org, UK

Shahla Seifi
Social Responsibility Research Network (SRRNet), SRRNet.org, UK

ISSN 2520-8772 ISSN 2520-8780 (electronic)
Approaches to Global Sustainability, Markets, and Governance
ISBN 978-981-99-2455-4 ISBN 978-981-99-2456-1 (eBook)
https://doi.org/10.1007/978-981-99-2456-1

This Springer imprint is published by the registered company Springer Nature Singapore Pte Ltd.
The registered company address is: 152 Beach Road, #21-01/04 Gateway East, Singapore 189721, Singapore

Introduction

The term sustainability has become one of the most significant in the current era. It seems to be ubiquitous among academics, politicians, business leaders, media personnel, and even the general public. Sustainability therefore has become one of the most important issues for businesses, governments, and society at large. Increasingly, it features in all planning for future activity. The topic of sustainability is under much debate as to what it actually is and how it can be achieved, but it is completely certain that the resources of the planet are fixed in quantity, and once used, cannot be reused except through being reused in one form or another. This is particularly true of the mineral resources of the planet. These are finite in quantity, and once fully extracted, extra quantities are no longer available for future use.

Similarly, the recent wars have shown us that food availability can be a problem which is subject to geopolitics and access to necessary food can be difficult when events disrupt global supply chains. Equally global events such as the pandemic have also shown us the fragility of supply chains. Unfortunately, the world has become largely dependent upon a small number of crops—such as wheat, maize, rice, barley, soybeans, potatoes, and palm oil—grown in a small number of regions. As demand for these crops has grown yields have been increased by the use of chemical fertilisers, pesticides, and herbicides which damage the land and degrade yields in the longer term. Additionally these chemicals find their way into the water supply and eventually damage marine ecosystems, disrupting other sources of food.

While demand for food is increasing as the population rises the available land to grow such food is shrinking, partly as it becomes built over. To compensate additional natural habitat is destroyed to create more space for crops which means that animal habitats and sources of other foods are reduced and ultimately lost. As if this was not enough the changes to climate we are experiencing in making crops less productive in normal locations as drought becomes more prevalent and frequent. We have already seen how irrigation and extensive overuse of water sources has led to the drying up of large sources such as the Aral Sea and Lake Chad. Similarly aquifers are becoming dry. Efforts are being put into developing and growing more drought-resistant varieties of crops but even so yields can be expected to reduce, even as demand increases. And these effects can only be expected to increase and become more severe as climate change continues to gather pace.

We have all heard about floods and fires and how they have an effect upon the landscape. Equally we have all heard that sea levels are rising. This takes place by the ice sheets at the poles melting but also by the expansion of water as it warms. Already sea levels have risen by over 20 cm from pre-industrial levels and are expected to rise by another 30 cm before the end of the century. When we talk about the sea level rising by 1 mm per year this does not seem worth bothering about but when we say that this is half a metre over time then we can see its significance. Additionally factors such as soil erosion and extraction of oil and groundwater have meant that the effect is much greater in some parts of the world. One obvious effect is mass migration as land disappears, and the subsequent problems which are caused by this. There is another effect of climate change however which is becoming obvious but has not yet been brought to significance. This is the frequency of landslips caused by excessive rain and storms. Such landslips might not matter too much except when they interrupt transport and destroy infrastructure. Such landslips destroy reads and rail line or interrupt transport until they can be repaired. More extreme cases have been known to cause extensive flooding and the destruction of dams—causing extreme devastation as the water is released and flows downhill, destroying everything in its past, and causing significant loss of life. Unfortunately we can only expect such events to become more severe and more frequent.

Although achieving net zero has become a mantra of the Western world it seems that little thought has been given to getting there and the consequences. Actions have been outlined which need to be undertaken such as changing all cars to electric and insulating houses. The reality however is that as climate changes the requirements of housing, and even transport, are also changing. The changing climate will mean more heat, more cold, more rain, more flooding in certain areas and possibly all of these in one particular area. So our hoses need to be adapted to the extremes of climate which will occur. Additionally there has been much building—of houses and other buildings—in what is the flood plain of rivers which are likely to flood and in the path of water if dams collapse. We have already seen the effects of these in certain areas and the devastation and loss of life which ensues. We also know that sea levels are rising and this will be up to a metre by the end of the century. When this applies to islands in the pacific then we are not too bothered but when it starts to affect—as it will—the coast in California and villages in the UK then it becomes real and the consequences become apparent.

We need to be realistic and recognise that the global supply chains for food are no longer possible and we will need to change our diet to that of our grandparents—eating only food which can be grown locally. Similarly we must accept that private vehicles for transport are no longer realistic possibilities and we will need to rely on public transport—which has been devastated in the last 50 years by lack of demand and will need to be reinstated. Similarly holidays to far-flung parts of the world will become more difficult. In general life will become less comfortable and it is by no means certain that people are ready to accept this. One additional factor is that this approach must be global and must apply to everyone and this will have political implications as we will discuss later.

Thus we can all see that there are changes happening in the world, and arguably, it is entering an era in which, dealing with the consequences of climate change and environmental problems while also recovering from the global pandemic. The United Nations however has declared that the climate is the greatest problem facing the world. This can be extended to an overarching concern with sustainability. There have been many who have recounted the problems but much fewer who have proposed solutions which can be adopted at a local level. What has been proposed can generally be found in selected journals such as New Scientist, Technological Sustainability, Social Responsibility Journal, and Sustainability. But the field is vast as these individual papers published only deal with individual issues. This book is intended to add to the discourse by its various contributions.

This book therefore sets out to examine aspects of the future and our need for sustainability. Thus we focus upon the current situation with respect to CSR while also considering the extent to which the focus has changed so much that we need to think about new approaches to our understanding of corporate behaviour and differing effects in practice. The international origins of the contributors to this volume make this an original contribution taking some of the best ideas from around the world.

This approach is based on the tradition of the Social Responsibility Research Network (a worldwide body of scholars with a membership of several thousand), which in its 20-year history has sought to broaden the discourse and to treat all research as interrelated and relevant to the business. This tradition has always been to explore the subject widely and to seek relevant solutions, while also sharing best practices. This book is based upon some of the contributions from the Network at our recent conference and shows both commonality and diversity in approaches and effects. Accordingly, the audience of this book is wide and includes but is not limited to academics in different fields including politics, industrial engineering, business, science, and social science. This is a multidisciplinary topic.

David Crowther
Shahla Seifi

Contents

Contributors

Fayez Albadri Middle East University, Amman, Jordan

Najwa Ashal Middle East University, Amman, Jordan

Ambareen Beebeejaun University of Mauritics, Réduit, Mauritius

Khoyratty Bushra Université Des Mascareignes, Beau Bassin-Rose Hill, Mauritius

David Crowther University of Bedfordshire, Luton, UK;
Social Responsibility Research Network (SRRNet), SRRNet.org, UK

Maria Costa Polytechnic Institute of Coimbra, Coimbra Business School, ISCAC, Coimbra, Portugal

Marcia Juliana d'Angelo Fucape Business School, Vitória, Espírito Santo, Brazil

Bheekharry Normada Devi Université Des Mascareignes, Beau Bassin-Rose Hill, Mauritius

Cristina Góis Polytechnic Institute of Coimbra, Coimbra Business School, ISCAC, Coimbra, Portugal

Srushti Govilkar MICA—The School of Ideas, Ahmedabad, India

Kritika Jaiswal Greenstraw, Ahmedabad, India

Vimi Neeroo Lockmun-Bissessur Université des Mascareignes, Beau Bassin-Rose Hill, Mauritius

Chris McLean University of Manchester, Manchester, UK

José Lázaro Oliveira Nunes Fucape Business School, Vitória, Espírito Santo, Brazil

Flávio Oliveira CSG/SOCIUS—ISEG—Universidade de Lisboa, Lisbon, Portugal

Swaleha Peeroo Université des Mascareignes, Beau Bassin-Rose Hill, Mauritius

Dineshwar Ramdhony University of Mauritius, Moka, Mauritius

Raysa Geaquinto Rocha NECE Research Center, University of Beira Interior, Covilhã, Portugal

Martin Samy Kedge Business School, Talence, France

Maria João Santos CSG/SOCIUS—ISEG—Universidade de Lisboa, Lisbon, Portugal

Aatman Shukla Kochar and Co, New Delhi, India

Ruchi Tewari MICA—The School of Ideas, Ahmedabad, India

Subrun Veerunjaysingh Université Des Mascareignes, Rose-Hill, Mauritius

Clara Viseu Polytechnic Institute of Coimbra, Coimbra Business School, ISCAC, Coimbra, Portugal

Part I
Changing Practice

Chapter 1
Fostering Digital Sustainability in Higher Education Through e-Learning Concepts

Khoyratty Bushra and Bheekharry Normada Devi

Abstract The aim of this paper is to explore the potentiality of digital learning practices in bringing sustainable development to a sustainable education system. Scholars have deeply researched on topics of e-learning related to curricula development, innovative pedagogical practices, platform and educational tools, and changing roles of lecturers and students, among others. However, very few studies have been conducted in the field of sustainability and digital education.

Digital Learning has proven to be the only reliable alternative as a replacement for traditional face-to-face learning methods across the globe when we have been struck by the COVID-19 pandemic. Academic institutions worldwide have invested heavily in e-learning practices and most of the courses offered in traditional classroom mode have been converted into e-learning mode. The success of digital learning initiatives needs to be ensured to make it a sustainable mode of learning. Based on an extensive literature review, a proposed theoretical model has been developed and tested empirically. In this paper, the technology acceptance model (TAM) concept was used to construct a theoretical model of e-learning acceptability and we present the impact that the use of an online learning platform has had in the public university in Mauritius, namely Université des Mascareignes. We showed the positive impact this education paradigm had on the acceptance of students and proposed factors contributing to a more sustainable learning and academic performance.

The model identifies some success determinants and relates them to different success measures, including learning and academic performance.

Keywords e-Learning acceptance · Technology acceptance model (TAM) · Sustainability in higher education · Educational platform · Digital learning · Learning culture

K. Bushra (✉) · B. N. Devi
Université Des Mascareignes, Beau Bassin-Rose Hill, Mauritius

D. Crowther and S. Seifi (eds.), *Preparing for a Sustainable Future*,
Approaches to Global Sustainability, Markets, and Governance,
https://doi.org/10.1007/978-981-99-2456-1_1

1.1 Introduction

The debate over sustainability and management education has its roots in international initiatives sponsored by the United Nations (UN), as the 2005-launched Global Compact aims to promote the adoption of CSR and sustainability policies. The 17 Sustainable Development Goals serve as Agenda 2030's main tenet, and they are connected to the Global Compact (UN Global Compact, 2020). The UN has also placed a strong emphasis on education to advance sustainable development, designating the past ten years as the "Decade of Education for Sustainable Development" (2005–2014) and establishing a global mandate to incorporate sustainability principles, values, and practices into education, research, operation, and evaluation (Buckler & Creech, 2014). The UNESCO Global Action Programme on Education for Sustainable Development (2015–2019) and the UN have both been launched. Higher education institutions across the globe are at a determining point. Unplanned changes as we all know, the COVID-19 and technological advancement have reshaped not only the mode of learning delivery, the learning environment, and learning platforms but also the overall operation in Higher Education. Quality education (Goal 4) aims not only to provide inclusive and equitable education, but also to promote learning opportunities (Alevar et al., 2019). In order to provide inclusive and equitable education, it is important to understand the inter-relationship among the 17 SDGs as demonstrated by Schuller et al. (2004), McMahon (2010), Psacharopoulos and Patrinos (2018), and Boeren (2019). The findings brought forward by these authors extrapolate on the benefits of how a proper education system facilitates the attainment of higher skills to secure prospective jobs, entertain the proliferation of talent development for a high-performing workforce, appreciate better health, engage in communities and social relationship, and focusing on active citizenship hence resulting in a sustainable world. Higher education institutions are increasingly recognizing the importance of learning culture as a core factor for students' sustainable learning and development (Eid & Nuhu, 2009), because learning culture directly influences or inhibits the quality of learning (Szulanski, 1996). Hence, developing a positive learning culture is a dominant theme in the strategic plans of higher education institutions. Basically, the traditional teaching paradigm is being reshaped by two main driving forces: the continuous innovation in digital technology and unplanned forces as we all are aware of the COVID-19 pandemic.

The emergence of digital learning encompassing e-learning, e-learning platforms, and tools has unavoidably changed how higher education is delivered, forcing institutions to quickly create online teaching formats and approaches. While some organizations had already started employing virtual tools, for the majority of them, it was uncharted ground. As a result of the crisis, institutions have had to reconsider how they approach teaching and learning, opening the door to more flexible learning options, blended learning exploration, and the blending of synchronous and asynchronous learning. Simultaneously academic and non-academic students were encouraged to become more tech-savvy and foster pedagogical innovation.

We have devised an equilibrium between e-learning and sustainable education and come forward with the concept of Sustainable Digital Learning (SDL). The aim of this paper attempts to demonstrate how Sustainable Digital Learning can be reinforced in higher education. This paper is structured as follows: Sect. 1.1 introduction, Sect. 1.2 elaborates on the Literature Review, Sect. 1.3 describes the methodology used followed by Sect. 1.4 that is the Analysis, Sect. 1.5 Findings, and the conclusion in Sect. 1.6.

1.2 Literature Review

This section shows the different determining components in SDL. Our personal and professional lives are changing drastically because of digitalization. New business models and ways of living are made possible by the widespread use of new technology.

Abdullah and Ward (2016) defined e-learning, namely known as electronic learning as a medium of delivering an interconnected learning system via internet, intranet, and extranet to the end users. In the same article reference was made to the work done by Lee et al. (2011) whereby e-learning was defined as follows: an information system that can integrate a wide variety of instructional material (via audio, video, and text mediums) conveyed through e-mail, live chat sessions, online discussions, forums, quizzes, and assignments. Hence e-learning is an innovative and constructive learning technique designed to facilitate the transfer and sharing of knowledge from instructors/educators to students while providing a new learning style in a new interactive space. Yet, e-learning cannot be described to be efficient and effective if learners do not use it and if the facilitators do not first understand their role in using it or do not have the required skills and competency.

Since this study is based on the student's perspective, a determining factor in the efficiency and effectiveness of e-learning will be to identify the key influencing factors to promote the use of digital learning (Fig. 1.1).

SDL goes beyond the concept of e-learning and the lack of consensus regarding how to define SDL and its related terms, is not of concern, rather provides a chance for additional research and contributions, as well as for new fields of study and application, to emerge (Cullen, 2019). As specified by Tick (2019), the wide application, acceptance, and adoption of digital tools like smartphones and mobile devices, the overall transformation in the delivery of teaching and learning, and the transition of learning from offline to online platforms is undeniable, both in terms of individual learning and within organizational frameworks, from a traditional form through blended learning to a digital form. In higher education, both students and adults heavily rely on vast online educational platforms and self-directed study with their own smart and mobile devices. The value of SDL relies on understanding and promotion of behavioural changes in individuals (Setó-Pamies & Papaoikonomou, 2016) and in the case of this research, comprehending and evaluating the psychological behaviour of students are considered only.

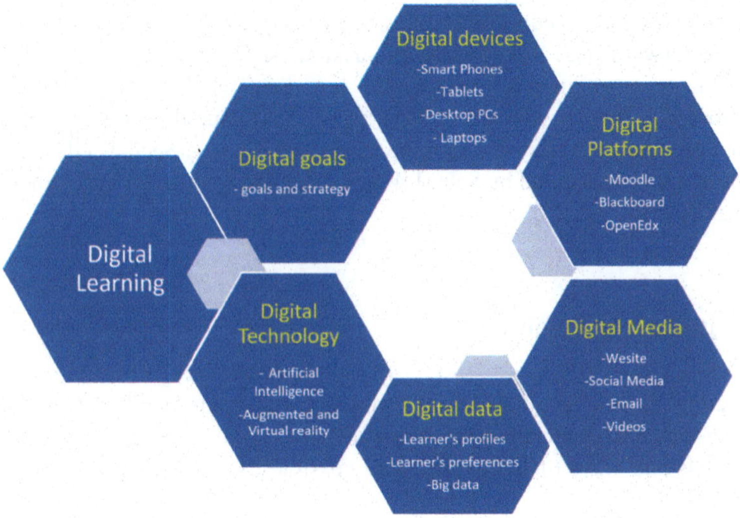

Fig. 1.1 Shows a framework showing interconnecting factors in digital learning (*Source* Elaboration on own literature review)

1.2.1 Sustainable Education

Sustainability can be interpreted from many different viewpoints, however, Tilbury (2007) defined sustainability as "the ability to continue an activity or a certain condition indefinitely". The bottom line for sustainability in education is dependent on the focus of designing and implementing sustainable forms of adapted and successful practices through educational development, leadership, and innovation (Coleman, 2013). Sustainable education is crucial for the development of individuals who represent the workforce of a nation. Countries investing in a sustainable education system and having a standard and continuous learning culture are going to promote growth and development, hence eradicating poverty, no room for hunger, fostering innovation, and building a resilient nation.

Sustainable education requires leaders of today and tomorrow to perform sustainable management and to train those who do not understand the origin and purpose of sustainability. In addition, curricula should be developed taking into consideration social and environmental responsibilities, as well as economic concerns (Wade & Parker, 2008). However, not only leadership is a key element in sustainable education but also the management and proliferation of knowledge. The twenty-first century is the period of digitization and millennials are prone to be continually connected with a peer group, social media, and entertainment, meaning they have a vast database of knowledge which can be used to restructure the learning system and customize it for a much more valuable output. The concept of sustainable education is outlined below, taking two major variables into consideration that are: Sustainable Digital Learning and Student Empowerment shown in Fig. 1.2.

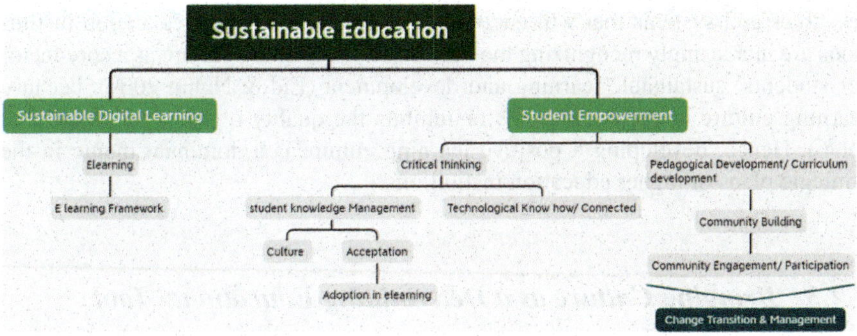

Fig. 1.2 Conceptual framework of SDL (*Source* Based on own elaboration)

1.2.2 Sustainable Digital Learning

As explained in the previous sections, higher education and sustainability are two closely related concepts. Higher education has a responsibility and an important role to play in redesigning education for sustainability in today's culture in addition to its two conventional roles of research and teaching (Sayaf et al., 2021). The ease of use and accessibility of technology have been able to connect people across the country and world while increasing at the same time the availability of resources (Agarwal et al., 2021; Sayaf et al., 2021). The COVID-19 period has given rise to a new pedagogical era (Agarwal et al., 2021) and this has been reinforced by the rapid adoption and use of educational technology (Sayaf et al., 2021). Higher education has included the integration of digital platforms in teaching delivery, assessment and evaluations, and other pedagogical functions and this is called digital learning (DL), (Sousa et al., 2022). Much research has been published on e-learning, e-learning tools and platforms, the role of facilitators, and the role of students, but not much has been done in digital learning.

DL also includes the usage of flipped classrooms, gamification, massive open online courses (MOOCs), digital learning, and hybrid learning were other instructional methodologies (Sousa et al., 2022). Digital technologies like augmented reality, graphic design, and a variety of interaction platforms were used to foster connections between lecturers and students during online classes. Following the pandemic, much research has recognized the need of incorporating internet platforms into teaching methods (Dost et al., 2020; Rajab et al., 2020).

In light of a digital sustainable educational programme, challenges to online education must be evaluated and further investigated (Dost et al., 2020; Mishra et al., 2020). Active student integration with technologies has already been shown to have many advantages. It incorporates problem-solving strategies that motivate students, improve their collaborative skills, train self-study skills, embrace diversity in a student population, and promote sustainable practices (Sousa et al., 2022). Accordingly, emerging instructional trends are effectively designed to better understand how the individuals, notably students, learn and how to develop educational resources and systems that will encourage this learning. Higher education institutions are increasingly recognizing the importance of learning culture as a core factor for students' sustainable learning and development (Eid & Nuhu, 2009), because learning culture directly influences or inhibits the quality of learning (Szulanski, 1996). Hence, developing a positive learning culture is a dominant theme in the strategic plans of higher education institutions.

1.2.3 Bringing Culture as a Determining Educational Tool

Learning culture has a positive impact on the knowledge-sharing behaviour of people in an organization (Eid & Nuhu, 2009). While the commitment to learning on the part of an organization and the learner should be required, which encourages learners to acquire new skills, to adopt new ways of practice, and to share knowledge (Johnston & Hawke, 2002). In some organizations this may be manifested by specific policies, corporate goals or documentation, and leadership style (Johnston & Hawke, 2002).

Conclusively, learning culture can be defined as "the existence of a set of attitudes, values and practices within an organization which support and encourage a continuing process of learning for the organization and/or its members" (Johnston & Hawke, 2002, p. 9). Furthermore, the learning culture is established by the combination of a commitment to learning, and a teamwork and open communication environment, in which the terms (encompassing formal and informal initiatives) of culture's attitudes, values, and practices are activated (Johnston & Hawke, 2002). This shows that learning culture is a phenomenon of the social constructivist context. The objective of research on cultural aspects of learning is to contribute to instructional design models in such a way that it is enhancing the knowledge of educational providers to provide learning materials which are adapted to the specific cultural requirements and differences of learners. It is thus in line with a broad range of empirical research attempting to elicit the preferences and requirements of learners in order to deal with their heterogeneity.

1.2.4 Emphasizing on Student Empowerment

Little is known about the actual process of empowering students for sustainability through higher education, even though empowerment for sustainability is thought to be a significant objective within sustainability-oriented educational programmes and policies (Tassone et al., 2017). Zimmerman (1995) defines empowerment as a process by which people gain mastery over issues of concern to them (Zimmerman, 1995). However, this process in education happens as disempowerment. Students acquire the skills necessary to meet their individual requirements in schools, such as learning, forming relationships with others, and receiving a diploma, as well as to collaborate with others (such as other students, instructors, and administrators) to accomplish group objectives (e.g., a safe and positive school environment; Prilleltensky et al., 2001). While individual and context-specific empowerment processes vary widely, empowered results feature intrapersonal, interpersonal, and behavioural components (Zimmerman, 1995). Students' senses of impact (or voice), competence, meaningfulness, and choice or self-determination are examples of intrapersonal outcomes (Thomas & Velthouse, 1990). Empowerment procedures are iterative and domain-specific, with greater power feeding into successive empowerment processes with no obvious beginning or conclusion (McWhirter & Wang, 1991). For example, empowerment at home may not be the same as empowerment at school (Zimmerman, 1995).

Frymier et al. (1996) believed that the fundamental ideas of empowerment in the workplace carried over to the sphere of education (while at the same time acknowledging the differences between the workplace and educational settings). A more promising paradigm, according to them, is to challenge faculty to foster student intrinsic motivation by removing obstacles that lead to feelings of helplessness and enhance personal significance and responsibility for their learning. They claimed that there are few extrinsic motivations that faculty can provide beyond course grades.

On the other hand, empowerment implies a shift of responsibility for learning from the institution or instructors to students. The role of instructors is not to simply create interesting activities in their courses that will engage students, but to create the conditions under which students are motivated to learn by the intrinsic nature of the design of the course. Empowerment tends to be a broader concept whereas engagement implies more of a task-specific focus. Empowered students are necessarily engaged in their work, but engaged students are not necessarily empowered. Next each of Thomas and Velthouse's (1990) four conditions for empowerment and how blended learning research supports each is reviewed. Where research evidence exists, blended learning is compared to fully face-to-face and fully online instruction with regard to empowerment.

1.2.5 TAM Model

The TAM model is based on a theoretical framework that serves to study how the perceived usefulness and the perceived ease of use of a new technology or a new service influence its acceptance. In other words, the TAM stipulates that the intentions to use a technology depend on its ease of use and the user's perceptions of usefulness. When the TAM was originally introduced by Davis (1989), it was a theory explaining the acceptance of technologies. Since then it has been used in various fields, including new technologies and services (Venkatesh, 2006). In the same year, King and He (2006) confirm that it is the most used model through scientific research to assess the acceptability of technologies. In the TAM model, the perceived usefulness refers to the level to which the user believes that using the technology will improve one's job performance, while the perceived ease of use refers to the degree of ease estimated by the potential user when using the technology. According to Davis

(1989), these two factors are considered the main determinants of user attitudes, and consequently directly influence their intentions of use. With scientific progress, this model has undergone two important changes namely, TAM2 (Venkatesh & Davis, 2000) and TAM3 (Venkatesh & Bala, 2008). In the TAM2, the two researchers tried to integrate external determinants of perceived usefulness through a longitudinal study of four organizations. However, in the TAM3 the authors tested eight theoretical models presenting the paradigm of technological acceptance in order to remove four determinants of the intention to accept a technology. The determinants in question are respectively: the expected performance, the expected effort, the social influence, and enabling conditions. Generally, these two extensions of the TAM model are seen as an important reinforcement for the work of Davis (1989). Ultimately, the situated acceptance approach invented by Bobillier-Chaumon and Dubois (2009) is based essentially on the principle of contextualization, which aims to analyse the contributions and determination of the limits of a technology in its context of use. This approach serves to determine the interest of a technology without detaching it from the specificities of the user's projects.

1.2.5.1 Applying the TAM Model in SDL

Figure 1.3 outlines the different components, consisting of student characteristics which are mainly driven by their culture, values, belief, attitude, perception, motivation, and technological knowledge and the adoption process which is directly influenced by student empowerment. All the elements play a vital role in SDL.

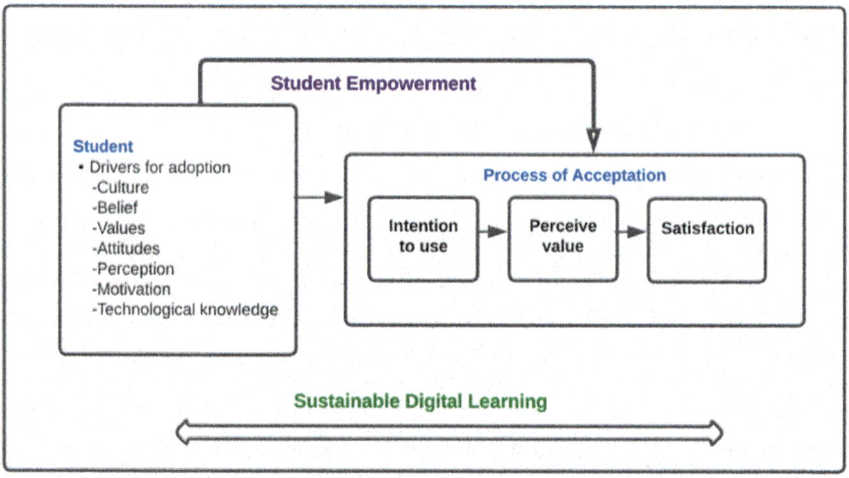

Fig. 1.3 Adapted TAM Model in SDL (*Source* Based on own elaboration)

1.2.5.2 Development of the Adapted TAM Model

With the help of extensive review of literature and application of different TAM models, this Adapted TAM model fits into a renewed theoretical perspective. Indeed, it considers the relationships between the three dimensions of acceptance namely: Intention to use, perceive value, and satisfaction. In addition, the model incorporates the influence of the characteristics of the individual learner and driving forces throughout the process of acceptance. In this process, user satisfaction appears as a major dimension in the extent to which acceptance is final only when users are satisfied with the technology (Dahab, 2001).

- Intention to use

Davis et al. (1989) added the behavioural intention dimension as they assumed that users would have an enhanced intention to use the technology if they had a high level of belief about the tool's usefulness, regardless of their attitudes. Various researches in different countries have shown that technology and quality of the Internet play an important role in determining students' intention to adopt e-learning courses (Piccoli et al., 2001; Webster & Hackley, 1997). A tool with user-friendly features such as learning and remembering tips with a few simple ideas and meaningful keywords require little work on the part of the users.

- Satisfaction

According to Wang (2003), satisfaction in an e-learning context measures responses and reactions about learner's emotions in relation to online pedagogical activities. The interface of the educational platform and its content are the two most frequently used metrics of satisfaction (Ramayah & Lee, 2012; Sun and al, 2008). According to Wang (2003), the proposed interface for the learner should be easy to use, understandable, and user-friendly. For the contents as well, it must be useful, comprehensive, high quality, and updated regularly. The current study proposes the following hypothesis to examine the correlation between satisfaction and integration of courses on the platform.

In the same context, user satisfaction appears as a major dimension so far as acceptance is not achieved until users are satisfied with the technology (Dahab, 2001). To measure satisfaction with the tool, the learner must first use it (Lee et al., 2003). Indeed, DeLone and McLean (2003) state that satisfaction is an evaluation of the technology resulting from its use. This predicts and influences customer satisfaction since according to the same authors, the most regular users are the most satisfied one.

Han and Sa (2021) suggest that student satisfaction has a positive effect on these users' behavioural intentions regarding future use of e-learning. Indeed, satisfaction or e-satisfaction is significant in the literature on online services, as this satisfaction influences the decision of whether to continue to use the service in question (Lin & Sun, 2009). Szymanski and Hise (2000) consider e-satisfaction as the judgement made by users on their experience globally online for a given period.

- Perceive value/usefulness

Lee et al. (2003) emphasize the importance for learners to perceive the usefulness of the educational platform and expected performance to achieve the acceptance of the tools. In addition, Perceived Usefulness was found to be the most significant variable in predicting the intention to use the online platform (Su & Tsai, 2013) while Saeed and Abdinnour-Helm (2008) found that Perceived Usefulness has an influence on the intention to use but was not the most influential factor. The current study proposes the following hypothesis to examine the correlation on perceived usefulness and ease of use of the learning platform.

1.2.6 Change and Transition Management

Hillman and Corkery (2010) suggest that educational management plays an important role in the development from distance to totally online programmes and online education cannot be viewed as a simple adjustment or a "tag on". That is, online programmes require a solid institutional infrastructure to support the programmes along with a willingness to collaborate and communicate among departments on campus. By necessity, an openness to question the "status quo" is essential (Hillman & Corkery, 2010, p. 473). Previous research has shown that drivers for the development of distance education many times are a consequence of a grassroots movement occurring at the department level (Hillman & Corkery, 2010). Other research also acknowledges the view that successful implementation of e-learning requires an enthusiastic grassroots movement driven by innovators. Establishing long-term stability, however, requires academic staff acceptance and engagement in combination with institutional structures in place to support the sustainability and mainstreaming of e-learning initiatives (MacKeogh & Fox, 2009; Roberts, 2008). Clearly, there are real obstacles encountered when implementing change such as distance learning when funding is tight and competing priorities are present. It also has to be acknowledged that while the support of senior management for change is essential, purely top down implementation strategies will not work in the traditional academic environment. The concerns and needs of academics and other stakeholders must also be addressed (MacKeogh & Fox, 2009, p. 152).

Implementing SDL will definitely require top management commitment and without any doubt there will be resistance to change, both from the facilitators and students. Managing this transition—from traditional to SDL will be challenging and the person in change, as a first step will need to identify Change Agents to push the project. Furthermore, clear and SMART digital goals should be communicated to everyone involved in the change process.

In addition, leadership is strongly associated with success; Löfström and Nevgi (2007), in their study of Helsinki University, conclude that commitment from department Heads is a critical factor in the development of web-based teaching, especially for maintaining high levels of pedagogical and technical support. This support level is needed to increase the number of teachers in web-based teaching. Similarly, Garrisson and Kanuka (2008) noted that in order to enable development, the senior leadership has to be strong and supports ideas and projects: "There must be a strategic selection of projects with accompanying resources, support and recognition" (Garrisson & Kanuka, 2008, p. 21). However, other researchers confirm that the implementation of technology in education is a question of educational management as well as management's relation to "internal factors such as resources, organizational culture, faculty readiness, anticipated degree of resistance, and the degree of variance from the status quo" (Roberts, 2008; Rovai & Downey, 2010). Similarly, Stoik (2001), on the basis of her study of integrating technology in education at a university, notes that strategic leadership and institutions have to be flexible and with the capacity to self-organize, adapt, and sustain themselves in an ever-changing environment.

1.2.7 Educational Context of Université Des Mascareignes (UDM)

UDM is the youngest public university in Mauritius, it is the result of the merger of two campuses: SD Campus Pamplemousses and Rose Hill Campus under the aegis of Université de Limoges in 2012. With its three faculties (Banking/Finance/Management; IT; Engineering), UDM delivers undergrade and post-grade courses and welcomes two types of learners: those who have just finished their higher school certificate and who enrol in full-time courses and more mature learners for continuous development that is working and studying in a parttime mode. Mauritian and international students (especially from Madagascar and Africa) are very prominent in full-time courses on the campus.

In its 10 years of experience, UDM has identified an atypical and emerging group of learners whose educational needs are constantly changing. In recent years and the impact of COVID-19, we have an audience that asks for more flexibility in the delivery of courses in terms of place and time, because often they must juggle between their student life, their professional life, and their family commitment. It is also important to note that the twenty-first-century learners being referred to as digital natives (Prensky, 2001) or emancipated learners (Littlejohn & Hood, 2018), our educational system is gradually moving from a "spoon-feeding" system to a more autonomous one with the positive impact of COVID-19.

The Université des Mascareignes, in line with the demand of Higher Education Commission (HEC) has devised an e-learning strategy. Its vision is to set up an efficient unity whose philosophy would be in line with that of the Minister of Education & Human Resources for Higher Education and Scientific Research, a knowledge community, based on scientific research and digital knowledge for the development of the Université des Mascareignes in the field of e-learning.

The fundamental objectives are to:

- Develop a digital strategy in close interaction with partners at the service of the academic mission of the UDM.
- Invest in educational innovation.
- Embody the Digital Institute of the Future both on the national and international level.
- On the other hand, the e-learning centre aims to promote higher education at UDM.
- It is a common service responsible for:
- Continuing teacher training in ICT and distance learning.
- The supervision of teachers in the production, co-production, or adaptation of online content and activities.
- The organization and management of distance learning for the benefit of learners.

This study emphasizes on the concept of building a sustainable e-learning framework. Since we have broadened the e-learning context to digital learning, the methodology developed is based on the core elements of student psychology, acceptance, and perceived ease of use. The following section outlines the methodology undertaken for the current study.

1.3 Methodology

To better understand student online behaviour, the study was conducted pre and after the total implementation of online learning at the Université des Mascareignes. Preliminary research was conducted in 2020 and the focus was to analyse (1) the ease of use of online tools, (2) the perceived intention of use the platforms, and (3) the level of satisfaction as per the new SDL Model introduced above. To compensate for an in-depth study, another research was done in March 2022. The major objective was to determine the restraining factors in accepting e-learning in Higher Education from the student's perspective. This descriptive research is aimed to have a broader view on the level of acceptance of e-learning, and what variables are hindering the implementation of e-learning which can subsequently, facilitate the Institutional Managers to devise proper digital strategies for the adopting of SDL.

1.3.1 Survey Conducted in 2020

The focus of the survey was to study the educational uses of digital technology during the period of confinement in March 2020 and the pedagogical issues imposed by distance education. It then analysed the specificities of distance education and the difficulties inherent with it in such a situation. The survey showed the practices that had been effectively put in place by lecturers at Udm, the difficulties they encountered in doing so and the positive lessons that could be drawn from this period.

1.3.2 Survey Conducted in 2022

An online questionnaire using google forms was designed for this purpose. It was open for the month March to April 2022. A series of multiple-choice and open-ended questions were set, to evaluate the digital platform put in place at Udm, while focusing on courses conducted in mixed mode during and after the pandemic. The questionnaire has four main constructs:

1. General demographic: age, gender, civil status, level of courses enrolled and preferences for blended or totally online mode of learning.
2. Platform assessment: determinants such as usage, preferences, motivations, perceptions, and level of satisfaction were measured.
3. Pedagogical (Teaching and learning materials) and the didactic aspects: we tried to get information on the appropriateness of the resources, relevance of the tools put at the service of the students, the way they learn, and the difficulties encountered.
4. Technology: Technological components relating to e-learning like course materials, evolution, and familiarization with the digital tools.

1.4 Key Findings

1.4.1 Students' Demography

Among the respondents registered in the year 2021–2022, there is a slightly higher proportion of male students (52%) than female students (48%) and their average age is 22 years with 72% of students aged between 22 and 25 years. More than 60% of students are not working, while 25.7% are working on a full-time basis and 13.7% are working part-time. Moreover, more than 40% of the respondents are final year students. 44.3% students still prefer a purely face-to-face classes, complemented, and supported by online resources.

The main reason for 39.4% of respondents (mostly final year students) choosing a distance education is that this mode of learning suited their way of learning and the second reason cited is that this form of learning suited students after a resumption of studies.

1.4.2 *Platform Assessment*

40.8% of students prefer Team's Interface as compared to Zoom and Moodle. This confirms the study done during the pandemic in March 2020 from which students claim their preference for Teams despite the Team's limitations in terms of functionalities and security (Table 1.1).

1.4.3 *Student's Appreciation in Terms of Perceived Usefulness and Ease of Use*

A strong and positive correlation was found between Perceived Ease of Use and Perceived Usefulness of the learning platform ($r = 0.683$, $n = 175$, $p < 0.01$).

This result confirms that the perceived usefulness of a distance education is strongly impacted by the degree of use expressed by students on the basis of their technological know-how in their field. Therefore, the perceived ease of use increases as perceived utility increases. This means that if students think that an e-learning platform is easy to use, they will use it systematically. This argument can be reinforced by the fact that students used e-learning during the COVID-19 health crisis without any prior preparation and our empirical results indicate that the perceived ease of use has a positive impact on student's continued use of the e-learning system. This conclusion is aligned with the literature on echnological acceptance by (Abdullah & Ward, 2016; Binyamin et al., 2019; Drueke et al, 2021; Han & Sa, 2021; Zogheib et al., 2015).

Table 1.1 How important was the Moodle platform for your course during and after the pandemic?

	Frequency	%. F
Essential	12	6.9
Not at all useful	17	14.3
Not very useful	25	9.7
Useful	121	69.1
Total	175	100.0%

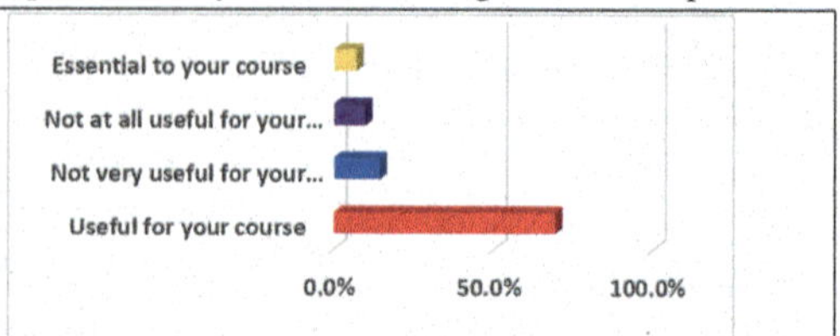

1.4.4 Student Satisfaction with the Integration of Their Courses on the Platform

There is a positive correlation between satisfaction and integration of courses on the platform and Perceived Ease of use ($r = 0.531$, $n = 175$, $p < 0.01$). In other words, it means that if the student is at ease with an online system, his level of satisfaction and productivity increases. The ease of use of the educational platform is a determining factor in the acceptance process, as it affects the intention to use and learner's satisfaction also (Ben Romdhane, 2013). This result is in line with those of Kwasi (2007) and Chang and Chang (2012) who found that the Perceived Ease of Use has a direct and positive impact and effect on the intention and attitude to use the system.

1.4.5 Student Engagement

Google Meet and Teams are defined by the authors (Ironsi, 2022) as a "synchronous learning tool for distant online programs".

The most widely used tools recorded in the survey 2022 during the e-learning sessions have been synchronous classes by sharing screens, chats, and e-mails. Microsoft Teams Rooms and their video conferencing features also provided a good basis for active interaction between lecturers and learners. Through them, effective synchronous work with students and their fuller engagement in the learning process is achieved. Research by many authors indicated the main possibilities used through these applications. Links to web-based classrooms are created by instructors and are integrated into learning courses in Moodle. This combines the use of these tools, with Moodle LMS offering much more comprehensive and complete support to students and faculty through a variety of activities and resources, but primarily asynchronous, and Google Meet and Microsoft Teams offering synchronous communication for lectures and seminars. Due to video conferencing functionality, live lectures or seminar sessions are held with students (Lewandowski, 2015) (Table 1.2).

1.4.6 Student Motivation

More than 75% of the student population seem to be motivated to use the e-learning platform. Nowadays the designs of platforms are doubling and shifting at extreme speed to capture attention, improve engagement, and increase learner satisfaction. The results of numerous studies show that the majority of students confirm that they are highly motivated, very active, and increasingly involved when their teachers use e-learning platforms in their courses. These results have been confirmed by several researchers and many research aimed at studying motivation in e-learning

Table 1.2 ICT tools used which were useful and engaging on the platform

Types	Frequency	% F.
Social Media	54	7.9
Virtual classes	82	12
Screen sharing	87	12.7
Visio-conferences	54	7.9
Audio conferences	64	9.3
Video Conferences	43	6.3
Uploading and downloading documents	67	9.8
Chat	90	13.1
Forum	54	7.9
Emails	90	13.1
Total	685	100.0%

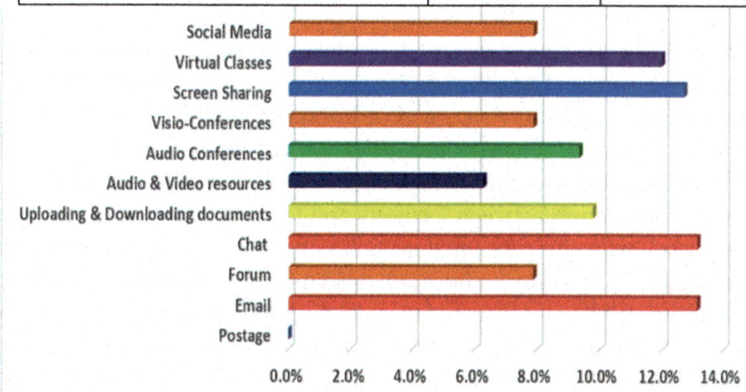

contexts. On the one hand, a study by Harandi (2015) shows that there is a significant relationship between e-learning and student motivation, and on the other hand, the use of platforms in pedagogical practices increases motivation among students to use the various tools integrated into the platform (Docq et al., 2010; Harandi, 2015). This is due to the fact that these platforms make available a series of toolkits and convey a large number of strategies to maintain and strengthen the motivation of students (Mignon & Closset, 2004) (Tables 1.3 and 1.4).

Table 1.3 What is your level of motivation when using Moodle as platform?

	Frequency	%. F
Not at all	16	9.1%
Little	25	14.3%
Medium	98	56.0%
Much	33	18.9%
Very much	3	1.7%
Total	175	100.0%

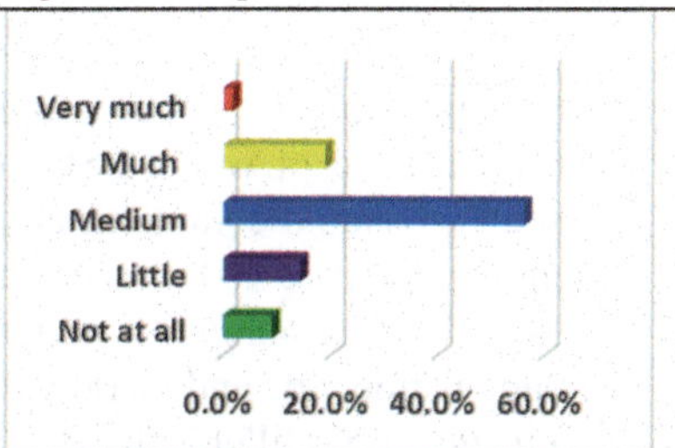

Table 1.4 How technology affects teaching?

	Frequency	%. F		Frequency	% F
I do not know	16	9.1	I do not know	11	6.3
Not at all	17	9.7	Not at all	4	2.3
Mostly no	22	12.6	Mostly no	12	6.9
Mostly yes	71	40.6	Mostly yes	92	52.6
Absolutely	49	28	Absolutely	21	12

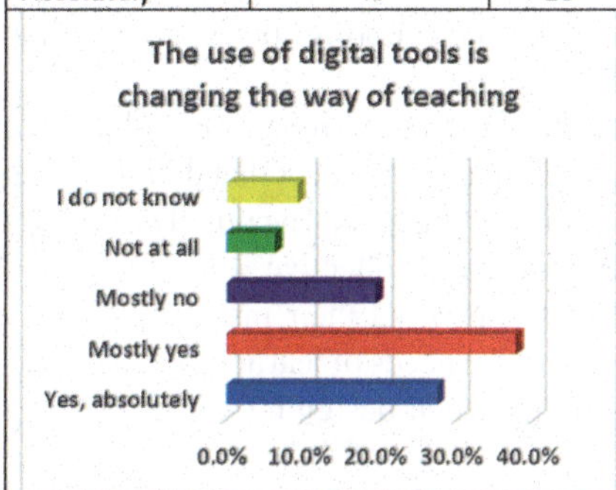

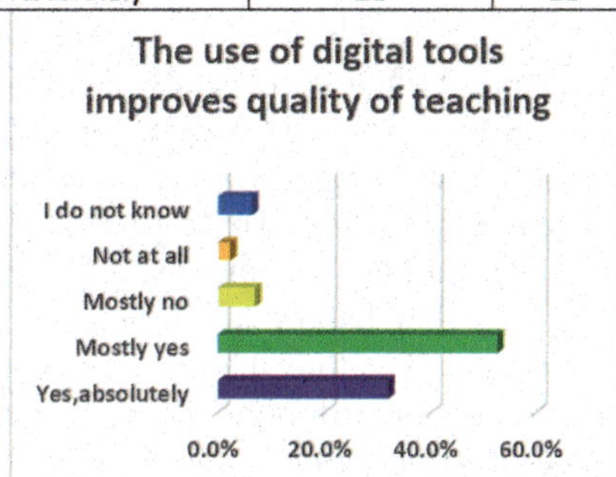

1.4.7 Technology

The result shows the effectiveness of ICT in teaching. More than 50% of the student population agree that technology has changed the way of teaching and also improves the quality of teaching. Several implications can be drawn from the results of this study. To begin with, when teaching with an online educational platform, more focus should be made on course design. Course quality may affect the continuance intention of students towards ongoing participation in online learning (Salloum et al., 2019). Thus, course information should be clear, comprehensive, and relevant. For example, lecturers should provide course outlines that include objectives, course materials, and time schedules. Moreover, rich multimedia content could be included and various online activities, such as discussion, could be designed in order to promote interactions with students (Yang et al., 2017).

1.5 Findings

1. Based on the analysis, we can assume that SDL is an emerging management and pedagogical concept, where the key players: management, institution, academics, and students should communicate to design and develop a continuous learning system—to prepare today and future generation for a poverty free, hunger free, social, and community welfare for a growing and innovative, sustainable economy nationwide and worldwide.
2. Participants are millennials—technological savvy and use technology for different purposes throughout the day, however a distinguished characteristic has been discovered. There is a huge acceptance of technology among this segment of learners however the extent of adoption of technology in learning is very low. Management of SDL should consider the depth of knowledge which millennials have and construct a distinguished curriculum with student support.
3. LMS platform has an effect on the level of acceptance of e-learning, throughout the analysis conducted in Table 1.1 and from the survey results, we understood that students prefer Ms Teams than Moodle. Hence, this study shows that the ease of use is a determining factor in accepting and adopting an education online platform bringing an overall satisfaction in learning.
4. Focus on developing and building an online student learning community. For instance, in an online learning community, all members expect that their learning needs will be satisfied by targeting a common learning goal. Eventually members of the learning community develop a common "collective consciousness", because they build relationships with one another and their instructors via an interface. Therefore, it is necessary to consider the needs of learners and the characteristics of each online learning community when designing online learning courses (Dede, 1996).

1.6 Conclusion

Technological development is an ongoing process and in the coming years, new and more innovative educational tools will be available on the market, simultaneously, more and more focus will be laid on sustainable education. Prior to these changes, HE institutions must devise ways to implement DL and at the same time make it sustainable. SDL is the future of higher education. However, not much research has been carried out in this field leaving a gap for further research in different areas: institution acceptance and management of DL, the proper education leadership style and characteristics, digital pedagogical design and curricula development, research in mass customized, creative and participative education platforms models.

According to the sustainable development education initiative, our findings may contribute to the existing education plans and future development either in Mauritius or other countries with the same education system and why not developed countries. We should also consider that the degree of technology and DL adoption will not be the same in each country as each and everyone has a different culture, which determines their motivation, attitude, and perception.

It is worth noting that students not only attach great importance to coping with challenges but also to carrying on a dialogue with stakeholders through feedback. In a physical and online hybrid learning environment, students come to the centre of attention through multiple channels of participation (Holmén et al., 2021; Jamison et al., 2014). Students share their views through forums, chats, and emoji which illustrate their feeling and commitment. Technological diversity and transformation-oriented learning are emerging to deal with the changes occurring within the learning environment. Higher education should provide students with the possibility of a participatory and vivid change practice.

Students can become empowered when they are in courses that provide them with choice and flexibility, they are presented with an experience that is personally meaningful, they believe that they can succeed with a reasonable amount of effort, and if they know that what they do in a course will result in them succeeding (Frymier et al., 1996). It has been observed that students nowadays are more entertainment-oriented and prefer short and illustrative videos. We believe that out of this observation DL should be more entertaining and this can be a topic for future research.

1.7 Recommendation and Further Research

To understand the concept of SDL, we have conducted the research at UDM, however, to have a broader view and understanding, further research can be elaborated involving the characteristics of both emotional and psychological aspects of lecturers, the engagement of management, and for a more accurate analysis a comparative study with other HE institutions in the country could have been considered. SDL is a complex model where all key stakeholders should understand their roles in delivering quality learning practices and should also consider the changing demographics of students. With continuous technological innovations and the easy diffusion and acceptance of technological know-how, our audiences are more into virtual reality; constant research should be conducted with these transformation changes as the core ingredients and comparisons made to better understand SDL. However, to make digital education sustainable the higher education institutions need to define programmes of digital competencies development for students as well as teachers. They also need to invest in the creation of pedagogical models for digital education, including all learning cycles. For the sustainability of digital education, the higher education institutions need to change and adapt the curricula to meet the needs of new social and scientific challenges.

References

Abdullah, F., & Ward, R. (2016). Developing a General Extended Technology Acceptance Model for E-Learning (GETAMEL) by analysing commonly used external factors. *Computers in Human Behavior, 56*, 238–256.

Agarwal, P., Verma, A., & Malhotra, S. K. (2021). AN ANALYSIS ON PERCEIVED SERVICE QUALITY AND STUDENTS' SATISFACTION OF E-LEARNING DURING COVID-19 IN HIGHER EDUCATION. *The Online Journal of Distance Education and e-Learning, 9*(3), 341.

Agarwal, R., & Gopinath, M. G. (2021). *A proposal to end the COVID-19 pandemic.* International Monetary Fund.

Agrawal, R., Wankhede, V. A., & Nair, R. S. (2021). Analysis of drivers of digital learning in COVID-19 and post-COVID-19 scenario using an ISM approach. *Journal of The Institution of Engineers (India) Series B, 102*, 1143–1155

Avelar, A. B. A., da Silva-Oliveira, K. D., & da Silva Pereira, R. (2019). Education for advancing the implementation of the sustainable development goals: A systematic approach. *The International Journal of Management Education, 17*(3), 100322.

Ben Romdhane, E. (2013). Issues in learners' acceptance of e-learning tools: What are the dimensions and determinants at Tunisian universities? *International Journal of Technologies in Higher Education, 10*(1), 46–57.

Binyamin, S. S., Rutter, M. J., & Smith, S. (2019). Extending the technology acceptance model to understand students' use of learning management systems in saudi higher education. *International Journal of Emerging Technologies in Learning, 14*(3).

Bobillier Chaumon, M.-E. (2016). L'acceptation située des technologies dans et par l'activité: Premiers étayages pour une clinique de l'usage. *Psychologie Du Travail Et Des Organisations, 22*(1), 4–21. https://doi.org/10.1016/j.pto.2016.01.00

Bobillier-Chaumon, M.-É., & Dubois, M. (2009). L'adoption des technologies en situation professionnelle: Quelles articulations possibles entre acceptabilité et acceptation? *Le travail humain, 72*(4), 355–382.

Boeren, E. (2019). Understanding Sustainable Development Goal (SDG) 4 on "quality education" from micro, meso and macro perspectives. *International Review of Education, 65*(2), 277–294.

Buckler, C., & Creech, H. (2014). *Shaping the future we want: UN Decade of Education for Sustainable Development; Final report.* UNESCO.

Chang, C. C., Yan, C. F., & Tseng, J. S. (2012). Perceived convenience in an extended technology acceptance model: Mobile technology and English learning for college students. *Australasian Journal of Educational Technology, 28*, 809–826.

Chang, I. Y., & Chang, W. Y. (2012). The effect of student learning motivation on learning satisfaction. *International Journal of Organizational Innovation, 4*(3).

Coleman, G. (2013). Sustainability as a learning challenge. *Journal of Management Development.*

Cullen, J. G. (2019). Varieties of responsible management learning: A review, typology and research agenda. *Journal of Business Ethics*, 1–15.

Dahab, D. (2001). *Les déterminants de l'acceptation des technologies de l'information et de la communication par les utilisateurs finals: application à la messagerie électronique dans une organisation industrielle* (Doctoral dissertation, Toulouse 1).

Davis, F. D. (September 1989). Perceived usefulness, perceived ease of use, and user acceptance of information technology. *MIS Quarterly, 13*(3), 318–340.

Davis, F. D., Bagozzi, R. P., & Warshaw, P. R. (1989). User acceptance of computer technology: A comparison of two theoretical models. *Management Science, 35*(8), 982–1003. https://doi.org/10.1287/mnsc.35.8.982

Dede, C. (1996). The evolution of distance education: Emerging technologies and distributed learning. *American Journal of Distance Education, 10*(2), 4–36.

DeLone, W. H., & McLean, E. R. (2003). The DeLone and McLean model of information systems success: A tenyear update. *Journal of Management Information Systems, 19*(4), 9–30.

Docq, F., Lebrun, M., & Smidts, D. (2010). Analyse des effets de l'enseignement hybride à l'université: détermination de critères et d'indicateurs de valeurs ajoutées. *Revue internationale des technologies en pédagogie universitaire/International Journal of Technologies in Higher Education, 7*(3), 48–59.

Dost, S., Hossain, A., Shehab, M., Abdelwahed, A., & Al-Nusair, L. (2020a). Perceptions of medical students towards online teaching during the COVID-19 pandemic: A national cross-sectional survey of 2721 UK medical students. *BMJ Open, 10*(11), e042378.

Drueke, B., Mainz, V., Lemos, M., Wirtz, M. A., & Boecker, M. (2021). An evaluation of forced distance learning and teaching under pandemic conditions using the technology acceptance model. *Frontiers in Psychology, 12*, 701347. https://doi.org/10.3389/fpsyg.2021.701347

Eid, M., & Nuhu, N. A. (2009). *The impact of learning culture and information technology use on knowledge-sharing: A case of KFUPM.*

Figueiró, P. S., Neutzling, D. M., & Lessa, B. (2022). Education for sustainability in higher education institutions: A multi-perspective proposal with a focus on management education. *Journal of Cleaner Production, 339*, 130539.

Frymier, A. B., Shulman, G. M., & Houser, M. (1996). The development of a learner empowerment measure. *Communication Education, 45*(3), 181–199.

Garrison, R., & Kanuka, H. (2008). Changing distance education and changing organizational issues. In W. J. Bramble & S. Panda (Eds.), *Economics of distance and online learning. Theory, practice, and research* (pp. 13–25). Routledge.

Han, J. H., & Sa, H. J. (2021). Acceptance of and satisfaction with online educational classes through the technology acceptance model (TAM): The COVID-19 situation in Korea. *Asia Pacific Education Review*, 1–13.

Harandi, S. R. (2015). Effects of e-learning on students' motivation. *Procedia-Social and Behavioral Sciences, 181*, 423–430.

Hillman, S. J., & Corkery, M. G. (2010). University infrastructural needs and decisions in moving towards online delivery programmes. *Journal of Higher Education Policy and Management, 32*(5), 467–474.

Holmén, J., Adawi, T., & Holmberg, J. (2021). Student-led sustainability transformations: Employing realist evaluation to open the black box of learning in a challenge lab curriculum. *International Journal of Sustainability in Higher Education, 22*, 1–24.

Ironsi, C. S. (2022). Google meet as a synchronous language learning tool for emergency online distant learning during the COVID-19 pandemic: Perceptions of language instructors and preservice teachers. *Journal of Applied Research in Higher Education, 14*, 640–659.

Jamison, A., Kolmos, A., & Holgaard, J. E. (2014). Hybrid learning: An integrative approach to engineering education. *Journal of Engineering Education, 103*, 253–273.

Johnston, R., & Hawke, G. (2002). *Case studies of organisations with established learning cultures.* National Centre for Vocational Education Research.

King, W. R., & He, J. (2006). A meta-analysis of the technology acceptance model. *Information & Management, 43*(6), 740–755. https://doi.org/10.1016/j.im.2006.05.003

Kirk, C. M., Lewis, R. K., Brown, K., Karibo, B., & Park, E. (2016). The power of student empowerment: Measuring classroom predictors and individual indicators. *The Journal of Educational Research, 109*(6), 589–595.

Kwasi, A. G. (2007). Perceived usefulness, user involvement and behavioral intention: An empirical study of ERP implementation. *Computers in Human Behaviour, 23*, 1232–1248

Lee, H., Kim, J. W., Hackney, R. (2011). Knowledge hoarding and user acceptance of online discussion board systems in eLearning: A case study. *Computers in Human Behavior, 27*(4), 1431–1437.

Lee, J. S., Cho, H., & Gay, G. (2003). Technology acceptance and social networking in distance learning. *Jounal of Educational Technology and Society, 6*(2), 50–61. Récupéré du site de la revue: http://www.ifets.Info

Lewandowski, M. (2015). Creating virtual classrooms (using Google Hangouts) for improving language competency. *Language Issues: ESOL Journal, 26*, 37–42.

Lin, G. T., & Sun, C. C. (2009). Factors influencing satisfaction and loyalty in online shopping: An integrated model. *Online Information Review, 33*(3), 458–475.

Littlejohn, A., & Hood, N. (2018). *Reconceptualising learning in the digital age: The [un] democratising potential of MOOCs*. Springer.

Löfström, E., & Nevgi, A. (2007). From strategic planning to meaningful learning: diverse perspectives on the development of web-based teaching and learning in higher education. *British Journal of Educational Technology, 38*(2), 312–324.

MacKeogh, K., & Fox, S. (2009). Strategies for embedding e-learning in traditional universities: Drivers and barriers. *Electronic Journal of e-Learning, 7*(2), 147–154.

McMahon, W. W. (2010). The external benefits of education. In P. Peterson, E. Baker, & B. McGaw (Eds.), *International encyclopaedia of education* (3rd ed., pp. 260–271).

McWhirter, J. R., & Wang, J. Y. (1991). Activation of tyrosinase kinase and microfilament-binding functions of C-ABL by BCR sequences in BCR/ABL fusion proteins. *Molecular and Cellular Biology, 11*(3), 1553–1565.

Mignon, J., & Closset, J. L. (2004). Maintien et stratégies de renforcement de la motivation des étudiants dans l'enseignement à distance. In *21e congrès de l'Association internationale de pédagogie universitaire (AIPU)*.

Mishra, L., Gupta, T., & Shree, A. (2020). Online teaching-learning in higher education during lockdown period of COVID-19 pandemic. *International Journal of Educational Research Open, 1*, 100012.

Piccoli, G., Ahmad, R., & Ives, B. (2001). Web-based virtual learning environments: A research framework and a preliminary assessment of effectiveness in basic IT skills training. *MIS Quarterly*, 401–426.

Prensky, M. (2001). Digital natives, digital immigrants part 2: Do they really think differently? *On the Horizon*.

Prilleltensky, I., Nelson, G., & Peirson, L. (2001). The role of power and control in children's lives: An ecological analysis of pathways toward wellness, resilience and problems. *Journal of Community & Applied Social Psychology, 11*(2), 143–158.

Psacharopoulos, G., & Patrinos, H. (2018). Returns to investment in education: A decennial review of the global literature. *Education Economics, 26*(5), 445–458.

Rajab, M. H., Gazal, A. M., & Alkattan, K. (2020). Challenges to online medical education during the COVID-19 pandemic. *Cureus, 12*(7).

Ramayah, T., & Lee, J. W. C. (2012). System characteristics, satisfaction and e-learning usage: A structural equation model (SEM). *Turkish Online Journal of Educational Technology-TOJET, 11*(2), 196–206.

Roberts, C. (2008a). Implementing educational technology in higher education: A strategic approach. *The Journal of Educators Online, 5*(1), 1–16.

Rovai, A. P., & Downey, J. R. (2010). Why some distance education programs fail while others succeed in a global environment. *Internet and Higher Education, 13*, 141–147.

Saeed, K. A., & Abdinnour-Helm, S. (2008). Examining the effects of information system characteristics and perceived usefulness on post adoption usage of information systems. *Information & Management, 45*(6), 376–386.

Salloum, S. A., Alhamad, A. Q., Al-Emran, M., Monem, A. A., & Shaalan, K. (2019). Exploring students' acceptance of e-learning through the development of a comprehensive technology acceptance model. *IEEE Access, 7*, 128445–128462.

Sayaf, A. M., Alamri, M. M., Alqahtani, M. A., & Al-Rahmi, W. M. (2021). Information and communications technology used in higher education: An empirical study on digital learning as sustainability. *Sustainability, 13*, 7074.

Schuller, T., Preston, J., Hammond, C., Brassett-Grundy, A., & Bynner, J. (2004). *The benefts of learning: The impact of education on health, family life and social capital*. Routledge.

Setó-Pamies, D., & Papaoikonomou, E. (2016). A multi-level perspective for the integration of ethics, corporate social responsibility and sustainability (ECSRS) in management education. *Journal of Business Ethics, 136*(3), 523–538.

Sousa, M. J., Marôco, A. L., Gonçalves, S. P., Machado, A. D. B. (2022). Digital learning is an educational format towards sustainable education. *Sustainability, 14*(3), 1140.

Stoik, J. H. (2001). Technology's role in collaboration. Community College. *Journal of Research and Practice, 25*, 27–36.

Su, S., & Tsai, C., (2013). Extending the TAM model to explore the factors affecting intention to use telecare systems. *Journal of Computers, 8*(2), 525–532.

Szulanski, G. (1996). Exploring internal stickiness: Impediments to the transfer of best practice within the firm. *Strategic Management Journal, 17*(S2), 27–43.

Szymanski, D. M., & Hise, R. T. (2000). E-satisfaction: An initial examination. *Journal of Retailing, 76*(3), 309–322.

Tassone, V. C., Dik, G., & van Lingen, T. A. (2017). Empowerment for sustainability in higher education through the EYE learning tool. *International Journal of Sustainability in Higher Education, 18*(3), 341–358. https://doi.org/10.1108/IJSHE-12-2015-0209

Thomas, K. W., & Velthouse, B. A. (1990). Cognitive elements of empowerment: An "interpretive" model of intrinsic task motivation. *Academy of Management Review, 15*(4), 666–681.

Tick, A. (2019). An extended TAM model, for evaluating eLearning acceptance, digital learning and smart tool usage. *Acta Polytechnica Hungarica, 16*(9), 213–233.

Tilbury, D. (2007). Learning based change for sustainability: Perspectives and pathways. In *Social learning towards a sustainable world* (pp. 117–132).

United Nations Global. (2020). United Nations Global Compact official website. www.unglobalcompact.org (2020)

Venkatesh, V. (2006). Where to go from here? Thoughts on future directions for research on individual-level technology adoption with a focus on decision making. *Decision Sciences, 37*(4), 497–518. https://doi.org/10.1111/j.1540-5414.2006.00136.x

Venkatesh, V., & Bala, H. (2008). Technology acceptance model 3 and a research agenda on interventions. *Decision Sciences, 39*(2), 273–315. https://doi.org/10.1111/j.1540-5915.2008.00192.x

Venkatesh, V., & Davis, F. D. (2000). A theoretical extension of the technology acceptance model: Four longitudinal field studies. *Management Science, 46*(2), 186–204. https://doi.org/10.1287/mnsc.46.2.186.11926

Wade, R., & Parker, J. (2008). *EFA-ESD dialogue: Educating for a sustainable world*. UNESCO. Retrieved from http://unesdoc.unesco.org/images/0017/001780/178044e.pdf (accessed July 2022).

Wang, Y. S. (2003). Assessment of learner satisfaction with asynchronous electronic learning systems. *Information & Management, 41*(1), 75–86.

Webster, J., & Hackley, P. (1997). Teaching effectiveness in technology-mediated distance learning. *Academy of Management Journal, 40*(6), 1282–1309.

Yang, M., Shao, Z., Liu, Q., & Liu, C. (2017). Understanding the quality factors that influence the continuance intention of students toward participation in MOOCs. *Educational Technology Research and Development, 65*(5), 1195–1214.

Zimmerman, M. A. (1995). Psychological empowerment: Issues and illustrations. *American Journal of Community Psychology, 23*(5), 581–599.

Zogheib, B., Rabaa'i, A., Zogheib, S., & Elsaheli, A. (2015). University student perceptions of technology use in mathematics learning. *Journal of Information Technology Education, 14*.

Chapter 2
Intrapreneurship: A Competent Method Toward Organizations' Sustainability

Najwa Ashal, David Crowther, and Fayez Albadri

Abstract This research argues that intrapreneurship is important to an organization's sustainability. This research discussed the definitions, dimensions, and importance of intrapreneurship. It spots the light on how intrapreneurship could be an efficient method for organizations' sustainability. Intrapreneurship could assist the organizations in adapting to internal and external environment changes, which in turn helps firms to innovate and revitalize their businesses. Intrapreneurship also could be a useful solution for guiding innovation advancement in different firms; thereby, it can resolve a range of complexities. Intrapreneurship is considered as an effective instrument to help the organization to revitalize, update, innovate, create new competencies, create new businesses, improve products, use resources efficiently, gain new opportunities, and access new markets. In other words, the successful employment of this process considered as an essential key to the organization's sustainability.

Keywords Corporate entrepreneurship · Entrepreneurial orientation · Innovativeness · Proactiveness · Risk-taking · Sustainability · New business venturing · Self-renewal · Competitive advantage

2.1 Introduction

In today's turbulent and uncertain business environment, COVID-19 pandemic, economic crises, and rapid technological change, business organizations have been continuously searching for new ways to sustain, generate, and exploit opportunities, manage innovation, and achieve competitive advantage (Aparicio et al., 2021a;

N. Ashal (✉) · F. Albadri
Middle East University, Amman, Jordan

D. Crowther
University of Bedfordshire, Luton, UK

Social Responsibility Research Network (SRRNet), SRRNet.org, UK

D. Crowther and S. Seifi (eds.), *Preparing for a Sustainable Future*,
Approaches to Global Sustainability, Markets, and Governance,
https://doi.org/10.1007/978-981-99-2456-1_2

Blanka, 2019; Woo, 2018). Intrapreneurship is considered as an effective method toward organization's sustainability. Nowadays with the changes in the economy and competition, organizations have to focus on the main three pillars of sustainability (economical, environmental, and social) to success and survive. Besides, nourishing the entrepreneurial environment could create an innovation culture that leads to improving organizational performance, and gaining a competitive advantage which is support the organizations' sustainability (Bilan et al., 2020).

Intrapreneurship (Corporate entrepreneurship) is considered as a process where the employees inside organizations detect and pursue opportunities using the resources they currently control (Antoncic & Hisrich, 2001). This process would lead to new business ventures, innovative activities, and orientations of modern technology, products, services, strategies, administrative techniques, and competitive postures (Woo, 2018). Intrapreneurs have the ability to transform the ideas into valuable results; therefore, they can seek and stimulate business venturing, product innovation, and strategic renewal activities. Organizational sustainability requires an active work system and processes with an efficient platform for future survival by focusing on how resources are used, renewing resources rather than consuming them for future business processes. Organizations have to focus on sustainability processes with a comprehensive and integrated practice context by integrating and employing factors of intrapreneurship and organizational learning. Intrapreneurship could assist the organizations to retrieve their vitality and nourish their innovation engine. With the current environmental and economic turbulence and the rapidly changing business, organizations working under the challenge of sustaining themselves. This research argues that intrapreneurship is considered as an efficient tool which assist organizations in enhancing the business's sustainability.

2.2 Literature Review

2.2.1 Introduction

The concept of intrapreneurship attracted the attention of many researchers since the beginning of the 1980s due to the mutual benefit for intrapreneurs and corporations if they worked together (Covin & Slevin, 1989; Miller, 1983; Rule & Irwin, 1988; Zahra, 1991). In 1985 Gifford Pinchot published a book emerged titled "Intrapreneuring: Why you don't have to leave the corporation to become an entrepreneur," he utilizes the term intrapreneur to characterize employees inside organizations who act as entrepreneurs. Then, during the years, researchers studied intrapreneurship from various aspects. Recently, intrapreneurship considered as a key subject of research in the fields of business and management (Ahmadpour Daryani & Karimi, 2018; Blanka 2019).

Corporate entrepreneurship, corporate venturing, and entrepreneurial activities are used as a synonym for intrapreneurship in the literature (Antoncic & Hisrich, 2001).

In the literature, intrapreneurship had been discussed from different views. The broadest view of intrapreneurship is that it is the entrepreneurship implemented by employees in the established organizations regardless of its size (Burgers & Covin, 2016). While the narrow view focusing on corporations and exclude the small organizations Kuratko et al. (1993). Intrapreneurship could be an aspirational goal of organizations to enhance the businesses' ability to innovate, renew, and revitalize (Hizarci-Payne, 2020; Urbano & Turró, 2013). In the same vein, intrapreneurship could assist organizations to compete and create added value in the ambiguous socioeconomic environments to achieve the sustainability (Lages et al., 2016). In the literature, there are different definitions of sustainability, as it is about achieving success today without affecting the needs of the future, so it is about meeting the needs of the present customers taking into consideration the needs of the next generations (Boudreau & Ramstad, 2005). Indeed, sustainability requires contemporary focus on the triple bottom line, which is social responsibility, environmental aspects, and economic profitability (De Clercq et al., 2018). Organizations have to make a balance between profit and its responsibility toward the society and environment, in which focusing on profit will not be sufficient (Dentchev et al., 2016). Sustainability has three main pillars: The environmental pillar which is about following the various practices to help the planet. Therefore, organizations could apply various practices to reduce their carbon footprints, and water usage. In addition, it has to take into consideration the way of packaging waste. As a result, these practices could positively impact the organization's financial situation by reducing spending and achieving efficiency. However, the social pillar focuses on the relationship between the organization and the other parties, such as the community, stakeholders, and employees, where the organization could have the approval and support of these parties, while the economic pillar (the governance pillar) concentrates on reasonable corporate governance. Where the board of directors and management meet the interest of the shareholders, value chains, community, and customers. In addition, this pillar clarifies that profit could not trump the environmental and social pillars.

Nowadays, there is an intensive competition between organizations to sell its goods and services; there is a scarcity of resources, in which organizations are facing a challenge of acquiring the needed resources and creating a demand for its products (Seifi & Crowther, 2017). Intrapreneurship considered as strategic tool to achieve the business success sustainability, as innovation, creativity, seeking opportunities, and renewal focuses on continuous development as products, markets to meet customers' needs (Srisathan et al., 2020).

2.2.2 Intrapreneurship Dimensions

Different authors explored different concepts and dimensions of intrapreneurship, such as, Miller, (1983) who stated that it is related to innovation in terms of (Innovation, Risk-taking, Proactiveness). While Covin and Slevin (1989) indicated that it is associated with the entrepreneurial posture in terms of (Risk-taking, Innovativeness, Proactiveness). Whereas Guth and Ginsberg (1990) and Zahra (1991) argued that it is related to the corporate entrepreneurship in terms of (internal innovation, venturing, strategic renewal). However, Lumpkin and Dess (1996) demonstrated that it is relevant to the entrepreneurial orientation dimensions (Autonomy, Innovativeness, Risk-taking, Proactiveness, Competitive aggressiveness), which is in line with Knight (1997) who stated that it is associated with entrepreneurial orientation in terms of (Innovativeness, Proactiveness). On the other hand, Antoncic and Hisrich (2001) and Antoncic (2007) argued that intrapreneurship has main four dimensions (proactiveness, new business ventures, innovativeness, and self-renewal) (Table 2.1).

There are similarities and differences between these works. Miller (1983) presented a seminal work in the intrapreneurship field. Covin and Slevin (1989), Lumpkin and Dess (1996) and Knight (1997) formulated the intrapreneurship dimensions based on Miller's work with different titles (entrepreneurial posture, and entrepreneurial orientation, respectively). They confirmed that entrepreneurship could be a phenomenon at the organizational level. Additionally, they suggested that entrepreneurial orientation might act as a signal of the entrepreneurship process by firms (Dao, 2018). Furthermore, Miller (1983) and Covin and Slevin (1989) focused on the innovation of products/markets, take some risk ventures, and to be proactive innovation to outperform the rivals. Besides, Lumpkin and Dess (1996) added two new dimensions (competitive aggressiveness and autonomy) to Miller's dimensions and they focused on enhancing the search for the opportunities and on challenging and outperforming the competitors.

On the other hand, Guth and Ginsberg (1990) and Zahra (1993) focused on the innovation and venturing activities which include the improvements in the different aspects of the organization such as the development of process, technology, market, administrative, and product. Moreover, they focused on strategic renewal which embraces the development of the organizational structure and the way of the combination of the resources. Finally, Antoncic and Hisrich (2001) combined between different dimensions from the works of the aforementioned authors.

Intrapreneurship has been discussed through two scales, the first scale named the *ENTRESCALE* used by Miller (1983), Covin and Slevin (1989), and Knight (1997) who stated that the ENTRESCALE reflects the generic direction of the organization toward the intrapreneurship and it also measures the proactive and innovative behaviors of an organization's management. The scale consists of two main dimensions: proactiveness and innovativeness. While the second scale named *corporate entrepreneurship* developed by (Zahra, 1991, 1993) who intended to measure the organization's engagement in the intrapreneurship activities using three measurements: "innovation, venturing, and corporate renewal activities." At the same context,

Table 2.1 Intrapreneurship dimensions

Author	Miller (1983)	Covin and Slevin (1989)	Guth and Ginsberg (1990) and Zahra (1993)	Lumpkin and Dess (1996) and Knight (1997)	Antoncic and Hisrich (2001)
Concept name	Entrepreneurship	Entrepreneurial posture	Corporate entrepreneurship	Entrepreneurial orientation	Corporate entrepreneurship
Dimensions	• "Innovation • Risk-taking • Proactiveness"	• "Risk-taking • Innovativeness • Proactiveness"	• "Internal innovation and venturing • Strategic renewal"	• "Competitive aggressiveness • Autonomy • Proactiveness • Risk-taking • Innovativeness"	• "Proactiveness • New business ventures • Self-renewal • Innovativeness"

there are two levels of intrapreneurship, individual level and organization level. Some authors explored intrapreneurship at the individual level such as (De Jong et al., 2011; Monsen et al., 2010). While other authors explored intrapreneurship at the organization level such as (Kreiser et al., 2011; Van der Sijde et al., 2013). Examining intrapreneurship by combining between the two scales would be comprehensive and the validity will be better Antoncic and Hisrich (2001). Regarding the organizational level, intrapreneurship could assist the organizations in enhancing the organizational performance, gaining competitive advantage, and supporting the organization's sustainability (Corrales-Garay et al., 2020; Kiyabo & Isaga, 2020). While according to the employee-level employing intrapreneurship may give the employees the opportunity to innovate, and it encourages them to be creative and get the self-achievement; as a result, it could enhance employees' satisfaction and engagement (Aina & Solikin, 2020; Gawke et al., 2017, 2019).

2.2.2.1 New Business Venturing

Various authors explored different dimensions related to the intrapreneurship, such as, (Covin & Slevin, 1989; Guth & Ginsberg, 1990; Knight, 1997; Lumpkin & Dess, 1996; Miller, 1983; Zahra, 1991), the new business venturing dimension appears in the studies by Zahra (1991, 1993); Antoncic and Hisrich (2001) who defined it as, "pursuing and entering new businesses related to the firm's current products or markets." Recent studies such as Baron (2006); Smith and Gregorio (2017) declared that intrapreneurs observe and examine the changes in the industry, market and customers then identify the links between these changes and their organization to propose opportunities for new business ventures. Likewise, Schmelter et al. (2010) shed light on the new business venturing generated inside the organization within its current markets. Launching new ventures could be done in two ways, internal corporate ventures or external corporate ventures. The new business with the internal corporate venture is generated and owned by the organization within its current organizational structure, while with the external corporate venture, new businesses are generated outside the organization in different ways such as acquiring or joint ventures (Covin & Miles, 2007; Kuratko & Morris, 2018). Intrapreneurs have the entrepreneurial behaviors which enable them to refine, propose, shape, and interpret the entrepreneurial opportunities of new businesses that may increase the organization competitiveness or performance (Kuratko & Audretsch, 2013).

2.2.2.2 Innovativeness

Innovativeness dimension has been discussed by the work of various authors in the intrapreneurship field, such as (Antoncic & Hisrich, 2001; Covin & Slevin, 1989; Knight, 1997; Lumpkin & Dess, 1996; Miller, 1983). Innovativeness is defined by Covin and Slevin (1989) and Antoncic and Hisrich (2001) as the origination of new products/services or technologies. Whereas Lumpkin and Dess (1996) set a broader definition of innovativeness as the organization's propensity to support and involve new experimentation, ideas, and creative process, leading to new products/services or new technology processes. Various authors examined the innovativeness and its importance, for example, Aktan and Bulut (2008), Schmelter et al. (2010), George and Marino (2011), Zehir et al. (2015), and Rank and Strenge (2018) spotlight on the organizations' emphasis on developing new products. Others such as Kaya (2006) and Aktan and Bulut (2008) focused on the organization's encouragement of the new innovative ideas, while Engelen et al. (2015) concentrated on the organization's creativity in the methods of operation. However, Schmelter et al. (2010), George and Marino (2011), Zehir et al. (2015), and Rank and Strenge (2018) focused on the organization's emphasis on Research and Development, and technological leadership. Other scholars such as Okun et al. (2020) and Poduška et al. (2020) examined the relationship between intrapreneurial climate and employee innovativeness and they found that the intrapreneurial climate encourages the employees to be innovative.

2.2.2.3 Self-Renewal

Self-renewal was defined by Antoncic and Hisrich (2001) as a reformulation of strategy, organizational change, and reorganization. Other authors such as Guth and Ginsberg (1990) and Zahra (1993) named it strategic renewal, which refers to renewal activities that create new wealth by combining different resources related to the business concept and the competitive approach. Whereas Aina and Solikin (2020) defined self-renewal as transformational practices of the organization which characterized by innovation and flexibility in order to respond to the needs and challenges of the market. From another perspective, Martín-Rojas et al. (2016) and Shafique and Kalyar (2018) defined self-renewal as an entrepreneurial effort which makes changes in the organization's business, hierarchy, and strategy; and these changes may transform the connections between the organization and its external environment. Self-renewal could enhance the response of the organization to the external environment (Kumar & Parveen, 2021). Renewal might assist the organization to get the ability and adaptability to effectively respond to the new changes (Boukamcha, 2019; Shu et al., 2019). Furthermore, renewal could serve the organization to grow, create new employment opportunities, improve its competitive advantage, and achieve prosperity (Schmelter et al., 2010; Schmitt et al., 2018).

2.2.2.4 Proactiveness

The proactiveness dimension also has been discussed by the work of various authors in the intrapreneurship field as (Antoncic & Hisrich, 2001; Covin & Slevin, 1989; Knight, 1997; Lumpkin & Dess, 1996; Miller, 1983). Proactiveness was defined by Covin and Slevin (1989) and Lumpkin and Dess (1996) as the organization seeking new opportunities and employing them ahead of the competitors; a proactive organization is a leader rather than a follower. From another point of view, Antoncic and Hisrich (2001) focused on their definition of the management orientation in seeking to promote competitiveness, and this may happen through initiative and risk-taking. Proactive organizations anticipate the environmental changes and act to these changes as a means to keep leading the competition (George & Marino, 2011; Kaya, 2006; Zehir et al., 2015). In the intrapreneurship literature, some authors tend to merge proactiveness and competitive aggressiveness, such as Covin and Slevin (1989), while others as Lumpkin and Dess (1996) indicated that there were differences between the two concepts, where the proactiveness was about how the organization reaction to market opportunities while competitive aggressiveness was about the organization reaction to competitors. In a different vein, there are various scholars such as Kaya (2006), Aktan and Bulut (2008), Schmelter et al. (2010), George and Marino (2011), and Zehir et al. (2015) who differentiate between proactiveness and risk-taking aspects, where proactiveness is about the organization's initiatives to get the opportunities to launch new things to stay ahead of its competitors. In contrast, risk propensity is about the degree of risk concerned with actions or projects and the organization's tendency to take these risks.

2.2.2.5 Risk-Taking

Risk-taking was considered as one of the intrapreneurship dimensions in the work of various authors such as (Covin & Slevin, 1989; Knight, 1997; Lumpkin & Dess, 1996; Miller, 1983). Risk-taking was defined by Miller and Friesen (1978) as the degree where the managers are willing to take risky and significant commitments to the resources. Likewise, Lumpkin and Dess (1996) argued that often, organizations with entrepreneurial orientation are characterized by risk-taking behavior, and they suggested that to measure risk-taking at the organization level, it could ask the managers about the organization's proclivity for careful versus bold projects and its tendency to engage in risky projects to achieve its aim. Innovativeness and proactiveness could be connected with the entrepreneurial behaviors, while risk-taking could be connected with the managerial attitude toward risk, in which it reveals the managers' desire to continue tracking the opportunities with uncertain or high outcomes (Eshima & Anderson, 2016).

2.2.3 Importance of Intrapreneurship

In the literature, there are various studies that have examined the relationship between intrapreneurship and other concepts and revealed different outcomes. The relationship between intrapreneurship and organizational performance has been examined by different studies employing different dimensions and different mediation variables. For example, Li et al. (2009), Felício et al. (2012), and Shan et al. (2016) investigated the effect of intrapreneurship in terms of (Innovativeness, Autonomy, Risk-taking, Proactiveness, Competitive Aggressiveness) on organizational performance; the results confirmed that intrapreneurship enhances organizational performance. Other researchers as Wang (2008), Simsek and Heavey (2011), and Ahmadpour Daryani and Karimi (2018) used learning orientation as a mediation variable between intrapreneurship and organizational performance with different dimensions of intrapreneurship (innovativeness, proactiveness, risk-taking, new business venturing, strategic renewal) and (Proactiveness, Aggressiveness, Risk-taking, Innovativeness), respectively. The studies found that learning orientation mediates the effect of intrapreneurship on organizational performance. On the other hand, Ahmed et al. (2020) examined the mediation role of employee engagement in the relationship between corporate entrepreneurship and business performance, the results revealed that employee engagement mediates the relationship between the two concepts. Moreover, a variety of studies such as Barrett and Weinstein (2015), Ambad and Wahab (2016), Umrani et al. (2018), Rezaei and Amin Fanak (2019), Aina and Solikin (2020), and Ziyae and Sadeghi (2020) found that intrapreneurship is a prominent factor which influences positively the organizational performance. From another perspective, various studies found a negative relationship between different intrapreneurship dimensions and organizational performance, for instance, Fitzsimmons et al. (2005), Ağca et al. (2012), and Jiménez-Barrionuevo et al. (2019) investigated the relationship between intrapreneurship in terms of (innovativeness, new business venturing, self-renewal, and proactiveness) and firm performance, employing a quantitative method to collect the data. The findings revealed that the self-renewal dimension has a negative relationship with the business performance. Similarly, other studies such as Hughes and Morgan (2007), Kraus et al. (2011), Boon et al. (2013), and Rahman et al. (2018) found a negative relationship between risk-taking dimension and business performance.

Intrapreneurship was deemed as a method that could help the organization to be proactive, refine the business concept continuously, define and meet the changes in customers' expectations/needs, and adopt innovative ways to achieve the competitive advantage (Urbano et al., 2013). Competitive advantage is defined by (Ferrell & Hartline, 2013, p. 16) as "something that the firm does better than its competitors that gives it an edge in serving customers' needs and/or maintaining mutually satisfying relationships with important stakeholders." This is in line with the studies by Sigalas and Economou (2013); Tareque and Islam (2020) who confirmed that the competitive advantage could be achieved by reasonable use of the organization's resources and capabilities in a way that achieves a special advantage compared with the competitors. From another perspective, Cantele and Zardini (2018) and Kryscynski et al. (2020) argued that the competitive advantage is linked with the value creation which could be achieved with implementation of the low-cost leadership or differentiation strategies. Moreover, organizations could achieve the competitive advantage in different ways, in which they could improve different aspects to be better than the competitors, such as products/services, R&D capabilities, managerial capabilities, profitability, and image (Singh et al., 2019). Furthermore, the essence of intrapreneurship is to promote enterprises to improve the business value and realize sustainable development (Ping et al., 2010; Yunis et al., 2018). Various studies examined the relationship between intrapreneurship and competitive advantage. For example, the studies generated by Zeebaree and Siron (2017), Mahmood and Arslan (2020), and Kiyabo and Isaga (2020) confirmed that intrapreneurship has a significant effect on competitive advantage. Other studies by Baruah and Ward (2014), Umrani et al. (2018), Boukamcha (2019), and Blanka (2019) found that intrapreneurship could help organizations to gain a competitive advantage.

In contrast, there are various studies which found that there was no significant relationship between intrapreneurship dimensions and competitive advantage, for example, Maryono et al. (2020) and Sadalia et al. (2020) studied the relationship between entrepreneurial orientation in terms of (innovativeness, proactiveness, and risk-taking) and competitive advantage; the findings revealed that there was no significant effect between risk-taking and the competitive advantage. Moreover, Stephen et al. (2019) examined the effect of entrepreneurial marketing dimensions on the competitive advantage of small and medium-size enterprises in Enugu state and found that proactiveness and risk-taking had no significant relationship with a competitive advantage.

Furthermore, there are various scholars such as Widya-Hastuti et al. (2016), Youssef et al. (2018), and Corrales-Garay et al. (2020) discussed the importance of applied the intrapreneurship dimensions to improve the organization's sustainability. In the same context, Brandi and Thomassen (2020) confirmed that organizations have to focus on prioritizing corporate entrepreneurship to promote the sustainability practices. Employing intrapreneurship assists organizations to beat on the sustainability challenges Dentchev et al. (2016).

In the literature, the light had been shed on the importance of intrapreneurship by exploring the organizational-level intrapreneurship and employee-level intrapreneurship (Gawke et al., 2019). On the organizational level, different scholars such as Wang (2008), Li et al. (2009), Simsek and Heavey (2011), Augusto Felício et al. (2012), Shan et al. (2016), and Ahmadpour Daryani and Karimi (2018) explored the role of intrapreneurship in enhancing the organizational performance. While, other scholars such as Baruah and Ward (2014), Zeebaree and Siron (2017), Umrani et al. (2018), Boukamcha (2019), Blanka (2019), Mahmood and Arslan (2020), and Kiyabo and Isaga (2020) found that intrapreneurship could help the organizations to gain competitive advantage.

From another perspective, the employee-level intrapreneurship had been studied and confirmed the importance of intrapreneurship with different concepts such as employee satisfaction (Ahmed et al., 2013; Auer Antoncic & Antoncic, 2011); employee engagement (Ahmed et al., 2020; Gawke et al., 2017; Özsungur, 2019). Intrapreneurship could give the employees the opportunity to innovate, and it encourages them to be creative and get the self-achievement (Aina & Solikin, 2020).

There are various studies that examined intrapreneurship comparing between developed and developing economies such as Alam et al. (2020) who articulated that intrapreneurship and innovation attained greater importance in the developed countries comparing with the developing countries. In the same vein, Aparicio et al. (2021b) conducted a research using panel data models for a sample of 43 countries between 2004 and 2012; they recommended future research to examine intrapreneurship, emphasizing on developing countries. Moreover, Guerrero and Urbano (2019) published an overview research about the effectiveness of technology transfer policies on entrepreneurial innovation across continents; with another research, Guerrero et al. (2020) compared the entrepreneurship process between developed and developing economies. They indicated that there was a debate in the literature about the status of entrepreneurial and intrapreneurial actions in the developing countries. In the same context, Aparicio et al. (2020) indicated that the quantity and quality of entrepreneurial activities are influenced by organizational and country levels; in developing countries, the entrepreneurial and intrapreneurial actions may be connected with survival purposes more than wealth creation and growth. Other studies such as Bhutta and Ali Shah (2015) and Alam et al. (2020) indicated that there was weak awareness about the importance of entrepreneurship in developing countries, so there is a need to future research to examine intrapreneurship in developing countries.

2.3 Discussion and Conclusion

This research discussed the definitions, dimensions, and importance of intrapreneurship. It discussed how the employment of intrapreneurship dimensions could improve the organizational performance, competitive advantage, employee engagement, and organization's sustainability.

Nowadays, with the rapid economic and environmental changes and severe competition with the open markets, organizations face a problem acquiring resources. The scarcity of resources in the world is considered a challenge for organizations to obtain the needed resources. There is severe competition for the limited quantities of different resources in the world. Thus, the transaction cost of acquiring these resources increases within time. Intrapreneurship is considered as an effective and comprehensive method to deal with this problem, where the entrepreneurial activities could assist the organization with the efficient use of the resources and operations which will lead to reduce the production cost. Also, it enhances the supply chain of the organization which will reduce the cost of acquiring the resources or distributing its goods. Moreover, the creative and innovative ideas could improve the status of the organization and expand its ability to gain new opportunities. This could be crucial to support the sustainability of organizations, countries, and world economy.

Intrapreneurship could be an aspirational goal of organizations to enhance the businesses ability to innovate, renew, and revitalize which in result supports organizations' sustainability. Organizations could consider intrapreneurship as an organizational strategy to sustain. Besides, intrapreneurship becomes a viable strategy to improve the organizational performance and sustained competitive advantage. Intrapreneurship was considered as a process in which the employees inside organizations detect and pursue opportunities using the resources they currently control or seek for opportunities to gain more resources as synergy and acquisition. This process would lead to new business ventures, innovative activities, and orientations of new technology, products, services, strategies, administrative techniques, and competitive postures (Gapp & Fisher, 2007; Woo, 2018).

As the severe competition imposes high pressure on the organization, it has to understand and visualize the customer's needs, and to be proactive in meeting these needs and desires. Intrapreneurs could seek opportunities which may help the organization to improve and sustain: first, they could seek opportunities for new business ventures in terms of internal corporate ventures with current industries (products/customers/markets) or external corporate ventures by acquiring or synergy, where the organization could exploit these opportunities to sustain. For example, if an organization makes a reasonable acquisition or synergy with other organizations, it will enhance the use of resources and achieve various benefits which may help it to sustain itself. Second, they could gain and employ new ideas, unique approaches, and creative processes in order to achieve new products, services, or production and technological processes. Third, intrapreneurs could revise the business concept, reorganize units/divisions, coordinate different businesses, and guide the organizational change, which in turn could help the organization adopt and sustain. Fourth, they could seek various initiatives to stay ahead of their competitors. Fifth, they could tend to employ risky behaviors to achieve a better return. For example, the entrepreneur Steve Jobs contributed to the Apple organization by developing innovative products, such as the iPod, iPhone, and iPad. In addition to his foundation the NEXT computer company, which generated an innovative operating system that was acquired from Apple to boost personal computers.

Moreover, intrapreneurs have the ability to transform the ideas into valuable results; therefore, they are not necessarily idea generators. Thus, they can seek and stimulate business venturing, product innovation, and strategic renewal activities. Employing intrapreneurship could encourage the managers and employees to seek different opportunities to improve the organizations' growth and sustainability. As Mimetic isomorphism theory declared that the organizations could imitate other organizations by adopting various behaviors of other organizations. On the other hand, intrapreneurs could respond to the stimuli that emerge from the environment of the organizations they work in. In addition, intrapreneurs could respond positively to stimuli arising from their social networks, as their friends in other organizations. These stimuli could be considered opportunities to develop entrepreneurial practices in organizations.

Indeed, organizations have to promote an organizational environment to the talented and creative employees who are continuously looking for proactive and innovative solutions besides better opportunities (Palazzeschi et al., 2018; Pandey et al., 2020). Furthermore, organizations have to create an innovative culture, enhance learning, and support employees in all levels to boost the intrapreneurs' opportunities to gain new ideas and take risks (Aguilar et al., 2019; Ben Arfi & Hikkerova, 2019).

Creating a learning environment and employing practices that encourage the employees to share their ideas and reward them could promote employees to think out of the box and provide creative ideas, where every employee will feel that his opinion is welcomed, and his solution could be the best to solve a problem. Additionally, if the employee has no incentive to share his ideas, s/he might remain in his/her comfort zone and receive orders and implement them only. In addition, the organizational environment where time is spent experimenting, thinking about innovative ideas, sharing knowledge, and receiving feedback could encourage and inspire employees for being entrepreneurial (Schenkel et al., 2019). The work environment might be considered one of the essential aspects supporting the development of intrapreneurship. Employees with intrapreneurship potential are more encouraged when they feel that the organization supports their actions (Piecuch & Szczygieł, 2021). The employees could be encouraged to be innovative, proactive, idea generators, and opportunities seizures when they work in environment which encourage learning such as: (1) they work in a comfortable atmosphere to talk about problems and share ideas, (2) they feel safe, and their jobs are secure, (3) their opinions are welcome, (4) they able to do their jobs in different ways, and (5) they have time for reflection. On the other hand, the leaders could have a key role in encouraging the employees to exchange knowledge and experience, which in turn could improve their entrepreneurial behavior. Furthermore, the brainstorming sessions, open discussion, and available time on non-routine tasks and activities could inspire the employees to be proactive, innovative, and idea generators. The organization could employ the leadership style which reinforces learning in terms of questioning and listening, encouraging discussion and dialogue, supporting the importance of knowledge transfer and reflection, and showing commitment to learning. Intrapreneurship might be a useful solution for guiding innovation advancement in different firms; thereby, it could resolve a range of complexities (Baruah & Ward, 2014; Palazzeschi et al., 2018; Pandey et al., 2020). Besides, intrapreneurial initiatives could improve the organizational performance and gain a competitive advantage (Gupta, 2016).

As a result of all these practices, intrapreneurship could enhance financial performance, competitiveness, and growth and create new competencies for the organization by using resources more efficiently and analyzing the internal and external environments continuously. In addition to making changes in the organization's system by running innovation, being proactive, and gaining opportunities. Intrapreneurs could present creative alternatives to the usage of resources. In sum, all of these improvements contribute effectively to the various pillars of sustainability and improve the organization's sustainability.

References

Ağca, V., Topal, Y., & Kaya, H. (2012). Linking intrapreneurship activities to multidimensional firm performance in Turkish manufacturing firms: An empirical study. *International Entrepreneurship and Management Journal, 8*(1), 15–33.

Aguilar, S., Vengrouskie, E. F., & Lloyd, R. A. (2019). Driving organizational innovation as a form of intrapreneurship within the context of small businesses. *Journal of Strategic Innovation and Sustainability, 14*(3), 25–28.

Ahmadpour Daryani, M., & Karimi, A. (2018). *Effect of corporate entrepreneurship on firm performance in Iranian ASMEs: The mediation role of knowledge creation and learning orientation.*

Ahmed, S. N., Rehman, A., & Amjad, M. (2013). Job satisfaction and intrapreneurship: The moderating effect of personality trait. *Journal of International Studies, 6*(1).

Ahmed, U., Umrani, W. A., Zaman, U., Rajput, S. M., & Aziz, T. (2020). Corporate entrepreneurship and business performance: The mediating role of employee engagement. *SAGE Open, 10*(4).

Aina, Q., & Solikin, I. (2020). Entrepreneurship and intrapreneurship: How supporting corporate performance. *Review of Integrative Business and Economics Research, 9*, 288–297.

Aktan, B., & Bulut, C. (2008). Financial performance impacts of corporate entrepreneurship in emerging markets: A case of Turkey. *European Journal of Economics, Finance and Administrative Sciences, 12*(8), 1530–2275.

Alam, M. Z., Kousar, S., Shabbir, A., & Kaleem, M. A. (2020). Personality traits and intrapreneurial behaviour. *Asia Pacific Journal of Innovation and Entrepreneurship, 14*(1), 31–46.

Ambad, S. N. A., & Wahab, K. A. (2016). The relationship between corporate entrepreneurship and firm performance: evidence from Malaysian large companies. *International Journal of Business and Society, 17*(2).

Antoncic, B. (2007). Intrapreneurship: A comparative structural equation modeling study. *Industrial Management & Data Systems, 107*(3), 309–325.

Antoncic, B., & Hisrich, R. D. (2001). Intrapreneurship: Construct refinement and cross-cultural validation. *Journal of Business Venturing, 16*(5), 495–527.

Aparicio, S., Audretsch, D., & Urbano, D. (2021a). Does entrepreneurship matter for inclusive growth? The role of social progress orientation. *Entrepreneurship Research Journal, 11*(4).

Aparicio, S., Audretsch, D., & Urbano, D. (2021b). Why is export-oriented entrepreneurship more prevalent in some countries than others? Contextual antecedents and economic consequences. *Journal of World Business, 56*(3).

Aparicio, S., Turro, A., & Noguera, M. (2020). Entrepreneurship and intrapreneurship in social, sustainable, and economic development: Opportunities and challenges for future research. *Sustainability*, 12(21).

Auer Antoncic, J., & Antoncic, B. (2011). Employee satisfaction, intrapreneurship and firm growth: A model. *Industrial Management & Data Systems, 111*(4), 589–607.

Augusto Felício, J., Ribeiro Soriano, D., Rodrigues, R., & Caldeirinha, V. R. (2012). The effect of intrapreneurship on corporate performance. *Management Decision, 50*(10), 1717–1738.

Baron, R. A. (2006). Opportunity recognition as pattern recognition: How entrepreneurs "connect the dots" to identify new business opportunities. *Academy of Management Perspectives, 20*(1), 104–119.

Barrett, H., & Weinstein, A. (2015). Corporate entrepreneurship, the marketing mix, and business performance. *In Proceedings of the 1997 Academy of Marketing Science (AMS) Annual Conference* (pp. 144–150).

Baruah, B., & Ward, A. (2014). Metamorphosis of intrapreneurship as an effective organizational strategy. *International Entrepreneurship and Management Journal, 11*(4), 811–822.

Ben Arfi, W., & Hikkerova, L. (2019). Corporate entrepreneurship, product innovation, and knowledge conversion: The role of digital platforms. *Small Business Economics, 56*(3), 1191–1204.

Bhutta, N. T., & Ali Shah, S. Z. (2015). Do behavioural biases impact corporate entrepreneurship, agency cost and firm performance: Evidence from developed and developing economies? *Pakistan Journal of Commerce and Social Sciences (PJCSS), 9*(3), 761–798.

Bilan, Y., Hussain, H. I., Haseeb, M., & Kot, S. (2020). Sustainability and economic performance: Role of organizational learning and innovation. *Engineering Economics, 31*(1), 93–103.

Blanka, C. (2019). An individual-level perspective on intrapreneurship: A review and ways forward. *Review of Managerial Science, 13*(5), 919–961.

Boon, J., Van der Klink, M., & Janssen, J. (2013). Fostering intrapreneurial competencies of employees in the education sector. *International Journal of Training and Development, 17*(3), 210–220.

Boudreau, J. W., & Ramstad, P. M. (2005). Talentship, talent segmentation, and sustainability: A new HR decision science paradigm for a new strategy definition. *Human Resource Management: Published in Cooperation with the School of Business Administration, the University of Michigan and in Alliance with the Society of Human Resources Management, 44*(2), 129–136.

Boukamcha, F. (2019). The effect of transformational leadership on corporate entrepreneurship in Tunisian SMEs. *Leadership & Organization Development Journal, 40*(3), 286–304.

Brandi, U., & Thomassen, M. L. (2020). Sustainable organizational learning and corporate entrepreneurship: a conceptual model of sustainability practices in organizations. *Journal of Workplace Learning.*

Burgers, J. H., & Covin, J. G. (2016). The contingent effects of differentiation and integration on corporate entrepreneurship. *Strategic Management Journal, 37*(3), 521–540.

Cantele, S., & Zardini, A. (2018). Is sustainability a competitive advantage for small businesses? An empirical analysis of possible mediators in the sustainability–financial performance relationship. *Journal of Cleaner Production, 182*, 166–176.

Corrales-Garay, D., Mora-Valentín, E.-M., & Ortiz-de-Urbina-Criado, M. (2020). Entrepreneurship through open data: An opportunity for sustainable development. *Sustainability, 12*(12).

Covin, J. G., & Miles, M. P. (2007). Strategic use of corporate venturing. *Entrepreneurship Theory and Practice, 31*(2), 183–207.

Covin, J. G., & Slevin, D. P. (1989). Strategic management of small firms in hostile and benign environments. *Strategic Management Journal, 10*(1), 75–87.

Dao, B. A. K. (2018). Danny Miller (1983) and the emergence of the entrepreneurial orientation (EO) construct. In *Foundational research in entrepreneurship studies.*

De Clercq, D., Thongpapanl, N., & Voronov, M. (2018). Sustainability in the face of institutional adversity: Market turbulence, network embeddedness, and innovative orientation. *Journal of Business Ethics, 148*(2), 437–455.

De Jong, J. P. J., Parker, S. K., Wennekers, S., & Wu, C. (2011). Corporate entrepreneurship at the individual level: Measurement and determinants. *EIM Research Reports. Zoetermeer: EIM, 11*(13), 3–27.

Dentchev, N., Baumgartner, R., Dieleman, H., Jóhannsdóttir, L., Jonker, J., Nyberg, T., Rauter, R., Rosano, M., Snihur, Y., & Tang, X. (2016). Embracing the variety of sustainable business models: social entrepreneurship, corporate intrapreneurship, creativity, innovation, and other approaches to sustainability challenges. *Journal of Cleaner Production.*

Engelen, A., Gupta, V., Strenger, L., & Brettel, M. (2015). Entrepreneurial orientation, firm performance, and the moderating role of transformational leadership behaviors. *Journal of Management, 41*(4), 1069–1097.

Eshima, Y., & Anderson, B. S. (2016). Firm growth, adaptive capability, and entrepreneurial orientation. *Strategic Management Journal, 38*(3), 770–779.

Felício, J. A., Rodrigues, R., & Caldeirinha, V. R. (2012). The effect of intrapreneurship on corporate performance. *Management Decision.*

Ferrell, O. C., & Hartline, M. (2013). *Marketing strategy, text and cases.* Cengage Learning.

Fitzsimmons, J. R., Douglas, E. J., Antoncic, B., & Hisrich, R. D. (2005). Intrapreneurship in Australian firms. *Journal of Management & Organization, 11*(1), 17–27.

Gapp, R., & Fisher, R. (2007). Developing an intrapreneur-led three-phase model of innovation. *International Journal of Entrepreneurial Behavior & Research, 13*(6), 330–348.

Gawke, J. C., Gorgievski, M. J., & Bakker, A. B. (2017). Employee intrapreneurship and work engagement: A latent change score approach. *Journal of Vocational Behavior, 100*, 88–100.

Gawke, J. C., Gorgievski, M. J., & Bakker, A. B. (2019). Measuring intrapreneurship at the individual level: Development and validation of the Employee Intrapreneurship Scale (EIS). *European Management Journal, 37*(6), 806–817.

George, B. A., & Marino, L. (2011). The epistemology of entrepreneurial orientation: Conceptual formation, modeling, and operationalization. *Entrepreneurship Theory and Practice, 35*(5), 989–1024.

Guerrero, M., Liñán, F., & Cáceres-Carrasco, F. R. (2020). The influence of ecosystems on the entrepreneurship process: A comparison across developed and developing economies. *Small Business Economics*.

Guerrero, M., & Urbano, D. (2019). Effectiveness of technology transfer policies and legislation in fostering entrepreneurial innovations across continents: An overview. *The Journal of Technology Transfer, 44*(5), 1347–1366.

Gupta, M. (2016). Intrapreneurship centric innovation: A step towards sustainable competitive advantage. *International Journal of Innovative Research and Development, 5*(2), 90–92.

Guth, W. D., & Ginsberg, A. (1990). Guest editors' introduction: Corporate entrepreneurship. *Strategic Management Journal*, 5–15.

Hizarci-Payne, A. K. (2020). Intrapreneurship. In *Encyclopedia of sustainable management*.

Hughes, M., & Morgan, R. E. (2007). Deconstructing the relationship between entrepreneurial orientation and business performance at the embryonic stage of firm growth. *Industrial Marketing Management, 36*(5), 651–661.

Jiménez-Barrionuevo, M. M., Molina, L. M., & García-Morales, V. J. (2019). Combined influence of absorptive capacity and corporate entrepreneurship on performance. *Sustainability, 11*(11).

Kaya, N. (2006). The impact of human resource management practices and corporate entrepreneurship on firm performance: Evidence from Turkish firms. *The International Journal of Human Resource Management, 17*(12), 2074–2090.

Kiyabo, K., & Isaga, N. (2020). Entrepreneurial orientation, competitive advantage, and SMEs' performance: Application of firm growth and personal wealth measures. *Journal of Innovation and Entrepreneurship, 9*(1).

Knight, G. A. (1997). Cross-cultural reliability and validity of a scale to measure firm entrepreneurial orientation. *Journal of Business Venturing, 12*(3), 213–225.

Kraus, S., Rigtering, J. P. C., Hughes, M., & Hosman, V. (2011). Entrepreneurial orientation and the business performance of SMEs: A quantitative study from the Netherlands. *Review of Managerial Science, 6*(2), 161–182.

Kreiser, P., Marino, L., Davis, J., Tang, Z. H. I., & Lee, C. (2011). Firm-level entrepreneurship: The role of proactiveness, innovativeness and strategic renewal in the creation and exploration of opportunities. *Journal of Developmental Entrepreneurship, 15*(02), 143–163.

Kryscynski, D., Coff, R., & Campbell, B. (2020). Charting a path between firm-specific incentives and human capital-based competitive advantage. *Strategic Management Journal, 42*(2), 386–412.

Kumar, S., & Parveen, R. (2021). Assessing the factors of employees' intrapreneurial behaviour: An Indian context. *World Journal of Entrepreneurship, Management and Sustainable Development*.

Kuratko, D. F., & Audretsch, D. B. (2013). Clarifying the domains of corporate entrepreneurship. *International Entrepreneurship and Management Journal, 9*(3), 323–335.

Kuratko, D. F., Hornsby, J. S., Naffziger, D. W., & Montagno, R. V. (1993). Implementing entrepreneurial thinking in established organizations. *SAM Advanced Management Journal, 58*(1), 28.

Kuratko, D. F., & Morris, M. H. (2018). Corporate entrepreneurship: A critical challenge for educators and researchers. *Entrepreneurship Education and Pedagogy, 1*(1), 42–60.

Lages, M., Marques, C. S., Ferreira, J. J. M., & Ferreira, F. A. F. (2016). Intrapreneurship and firm entrepreneurial orientation: Insights from the health care service industry. *International Entrepreneurship and Management Journal, 13*(3), 837–854.

Li, Y.-H., Huang, J.-W., & Tsai, M.-T. (2009). Entrepreneurial orientation and firm performance: The role of knowledge creation process. *Industrial Marketing Management, 38*(4), 440–449.

Lumpkin, G. T., & Dess, G. G. (1996). Clarifying the entrepreneurial orientation construct and linking it to performance. *Academy of Management Review, 21*(1), 135–172.

Mahmood, A., & Arslan, A. (2020). Ahead of the curve: Leveraging antecedents of corporate entrepreneurship to pull off competitive advantage. *Frontiers in Psychology, 11*, 531886.

Martín-Rojas, R., Fernández-Pérez, V., & García-Sánchez, E. (2016). Encouraging organizational performance through the influence of technological distinctive competencies on components of corporate entrepreneurship. *International Entrepreneurship and Management Journal, 13*(2), 397–426.

Maryono, Ghozali, I., Kusumawardhani, A., Prabantarikso, R. M., & Basbeth, F. (2020). Entrepreneurial orientation in government-owned bank: Do they improve competitive advantage? *International Journal of Financial Research, 11*(2).

Miller, D. (1983). The correlates of entrepreneurship in three types of firms. *Management Science, 29*(7), 770–791.

Miller, D., & Friesen, P. H. (1978). Archetypes of strategy formulation. *Management Science, 24*(9), 921–933.

Monsen, E., Patzelt, H., & Saxton, T. (2010). Beyond simple utility: Incentive design and trade-offs for corporate employee-entrepreneurs. *Entrepreneurship Theory and Practice, 34*(1), 105–130.

Okun, O., Arun, K., & Begec, S. (2020). Intrapreneurship and expectations restrictions. *Dimensión Empresarial, 18*(2).

Özsungur, F. (2019). Ethical leadership, intrapreneurship, service innovation performance and work engagement in chambers of commerce and industry. *Journal of Human Behavior in the Social Environment, 29*(8), 1059–1081.

Palazzeschi, L., Bucci, O., & Fabio, A. D. (2018). High entrepreneurship, leadership, and professionalism (HELP): A new resource for workers in the 21st century. *Frontiers in Psychology, 9*, 1480.

Pandey, J., Gupta, M., & Hassan, Y. (2020). Intrapreneurship to engage employees: Role of psychological capital. *Management Decision, 59*(6), 1525–1545.

Piecuch, T., & Szczygieł, E. (2021). The issue of intrapreneurship development in corporations. *Journal of International Studies, 14*(3).

Ping, W. L., Jie, J., Naiqiu, L., & Zhengzhong, X. (2010). Notice of retraction: A review and prospects of research on human resource management of intrapreneurship. *In 2010 IEEE International Conference on Advanced Management Science (ICAMS 2010)*, 3 (pp. 456–460).

Poduška, Z., Nedeljković, J., Nonić, D., Ratknić, T., Ratknić, M., & Živojinović, I. (2020). Intrapreneurial climate as momentum for fostering employee innovativeness in public forest enterprises. *Forest Policy and Economics*, 119.

Rahman, M. M., Chowdhury, A. A., & Dey, M. (2018). Relationship between risk-taking, capital regulation and bank performance: Empirical evidence from Bangladesh. *Eurasian Journal of Business and Economics, 11*(22), 29–57.

Rank, O. N., & Strenge, M. (2018). Entrepreneurial orientation as a driver of brokerage in external networks: Exploring the effects of risk taking, proactivity, and innovativeness. *Strategic Entrepreneurship Journal, 12*(4), 482–503.

Rezaei, R., & Amin Fanak, D. (2019). Mediation effect of organizational learning on the relationship between entrepreneurial orientation and organizational performance in the agriculture Jihad Organization of West Azerbaijan Province, Iran. *Journal of Agricultural Science and Technology, 21*(1), 37–50.

Rule, E. G., & Irwin, D. W. (1988). Fostering intrapreneurship: The new competitive edge. *The Journal of Business Strategy, 9*(3), 44–47.

Sadalia, I., Muharam, H., & Mulyana, A. (2020). The effects of entrepreneurial orientation and competitive advantage on the international market entry through financing. In *The future opportunities and challenges of business in digital era 4.0.* Routledge.

Schenkel, M. T., Farmer, S., & Maslyn, J. M. (2019). Process improvement in SMEs: The impact of harmonious passion for entrepreneurship, employee creative self-efficacy, and time spent innovating. *Journal of Small Business Strategy, 29*(1), 71–84.

Schmelter, R., Mauer, R., Börsch, C., & Brettel, M. (2010). Boosting corporate entrepreneurship through HRM practices: Evidence from German SMEs. *Human Resource Management, 49*(4), 715–741.

Schmitt, A., Raisch, S., & Volberda, H. W. (2018). Strategic renewal: Past research, theoretical tensions and future challenges. *International Journal of Management Reviews, 20*(1), 81–98.

Seifi, S., & Crowther, D. (2017). *Sustainability and resource depletion*. Emerald Publishing Limited.

Shafique, I., & Kalyar, M. (2018). Linking transformational leadership, absorptive capacity, and corporate entrepreneurship. *Administrative Sciences, 8*(2).

Shan, P., Song, M., & Ju, X. (2016). Entrepreneurial orientation and performance: Is innovation speed a missing link? *Journal of Business Research, 69*(2), 683–690.

Shu, C., De Clercq, D., Zhou, Y., & Liu, C. (2019). Government institutional support, entrepreneurial orientation, strategic renewal, and firm performance in transitional China. *International Journal of Entrepreneurial Behavior & Research.*

Sigalas, C., & Economou, V. P. (2013). Revisiting the concept of competitive advantage: Problems and fallacies arising from its conceptualization. *Journal of Strategy and Management.*

Simsek, Z., & Heavey, C. (2011). The mediating role of knowledge-based capital for corporate entrepreneurship effects on performance: A study of small- to medium-sized firms. *Strategic Entrepreneurship Journal, 5*(1), 81–100.

Singh, S. K., Chen, J., Del Giudice, M., & El-Kassar, A.-N. (2019). Environmental ethics, environmental performance, and competitive advantage: Role of environmental training. *Technological Forecasting and Social Change, 146*, 203–211.

Smith, K. G., & Gregorio, D. D. (2017). Bisociation, discovery, and the role of entrepreneurial action. *Strategic Entrepreneurship: Creating a New Mindset*, 127–150.

Srisathan, W., Ketkaew, C., & Naruetharadhol, P. (2020). The intervention of organizational sustainability in the effect of organizational culture on open innovation performance: A case of Thai and Chinese SMEs. *Cogent Business & Management, 7*(1), 1717408.

Stephen, O., Ireneus, N., & Moses, O. (2019). Entrepreneurial marketing practices and competitive advantage of small and medium size enterprises in Nigeria. *European Journal of Business and Innovation Research, 7*(30), 1–30.

Tareque, M. A., & Islam, N. (2020). Conceptualizing leadership behaviors and its relationship with competitive advantage: A perspective of RMG Bangladesh. *Journal of Leadership, Accountability and Ethics, 17*(6), 138–146.

Umrani, W. A., Kura, K. M., & Ahmed, U. (2018). Corporate entrepreneurship and business performance. *PSU Research Review, 2*(1), 59–80.

Urbano, D., Alvarez, C., & Turró, A. (2013). Organizational resources and intrapreneurial activities: An international study. *Management Decision, 51*(4), 854–870.

Urbano, D., & Turró, A. (2013). Conditioning factors for corporate entrepreneurship: An in(ex)ternal approach. *International Entrepreneurship and Management Journal, 9*(3), 379–396.

Van der Sijde, P., Veenker, S., & During, W. (2013). Intrapreneurship in SMEs: About the role of management and R&D. *European Journal of Business and Social Sciences, 1*(11), 24–30.

Wang, C. L. (2008). Entrepreneurial orientation, learning orientation, and firm performance. *Entrepreneurship Theory and Practice, 32*(4), 635–657.

Widya-Hastuti, A., Talib, N. B. A., Wong, K. Y., & Mardani, A. (2016). The role of intrapreneurship for sustainable innovation through process innovation in small and medium-sized enterprises: A conceptual framework. *International Journal of Economics and Financial Issues, 6*(3S).

Woo, H. R. (2018). Personality traits and intrapreneurship: The mediating effect of career adaptability. *Career Development International, 23*(2), 145–162.

Youssef, A. B., Boubaker, S., & Omri, A. (2018). Entrepreneurship and sustainability: The need for innovative and institutional solutions. *Technological Forecasting and Social Change, 129*, 232–241.

Yunis, M., Tarhini, A., & Kassar, A. (2018). The role of ICT and innovation in enhancing organizational performance: The catalysing effect of corporate entrepreneurship. *Journal of Business Research, 88*, 344–356.

Zahra, S. A. (1991). Predictors and financial outcomes of corporate entrepreneurship: An exploratory study. *Journal of Business Venturing, 6*(4), 259–285.

Zahra, S. A. (1993). Environment, corporate entrepreneurship, and financial performance: A taxonomic approach. *Journal of Business Venturing, 8*(4), 319–340.

Zeebaree, M. R. Y., & Siron, R. B. (2017). The impact of entrepreneurial orientation on competitive advantage moderated by financing support in SMEs. *International Review of Management and Marketing, 7*(1), 43–52.

Zehir, C., Can, E., & Karaboga, T. (2015). Linking entrepreneurial orientation to firm performance: The role of differentiation strategy and innovation performance. *Procedia - Social and Behavioral Sciences, 210*, 358–367.

Ziyae, B., & Sadeghi, H. (2020). Exploring the relationship between corporate entrepreneurship and firm performance: The mediating effect of strategic entrepreneurship. *Baltic Journal of Management, 16*(1), 113–133.

Chapter 3
Media Regulation in Mauritius: A Critical Analysis

Ambareen Beebeejaun

Abstract Indeed, media regulation confers substantive rights and liberties such as journalistic autonomy and freedom of expression and information. Enshrined in the Independent Broadcasting Authority (IBA) (Amendment) Act of Mauritius, voted on 30 November 2021, are explicit journalistic constraints which led to a public outcry, making the headlines of several national newspapers. The controversial highlight of this new amendment act is the excessive powers allotted to the IBA, being the media regulatory authority in Mauritius. Therefore, this paper aims to critically analyse the legal rules pertaining to the media in Mauritius, more specifically to discuss the powers of the Mauritian IBA and the salient features of the new amendments made to the IBA Act. To achieve this research objective, the black letter method will be adopted and a comparative analysis will be conducted on the legal and regulatory landscape of media in Mauritius against the corresponding legal provisions of South Africa.

Keywords Media law in Mauritius · Media freedom · Independent Broadcasting Act · Co-regulation in Media · Autocracy and media · Democracy and media · Media regulation in South Africa

3.1 Introduction

Press and media regulation has been introduced as a safeguard to prevent any abuse of power which may result from the freedom of expression as provided by Article 12(1) of the Constitution of Mauritius and by Article 10 of the European Convention on Human Rights. For decades, press companies have entertained differing audiences, becoming progressively profit-driven (Sindane, 2018). Hence, there came a time where various categories of content were diffused through social media outlets, newspapers, radios and television channels to seek new audiences and to keep the old

A. Beebeejaun (✉)
University of Mauritics, Réduit, Mauritius

D. Crowther and S. Seifi (eds.), *Preparing for a Sustainable Future*,
Approaches to Global Sustainability, Markets, and Governance,
https://doi.org/10.1007/978-981-99-2456-1_3

ones subscribed, notwithstanding their potential impact. To the extent that became necessary in the public interests, media regulatory bodies emerged to moderate such content before reaching their respective audiences (Ruckenstein & Turunen, 2019).

Likewise, Mauritius, being an island nation in the Indian Ocean, has its own press and media coverage. The successful evolution of the written press in the country has been witnessed by the increasing number of newspapers being published either on a daily basis either in the morning, afternoon or both or even on a weekly basis such as "Le Defi Quotidien", "L'Express Dimanche" or "Star" amongst others. In addition to written press, information is also broadcasted via audio-visual press. Historically, the first Mauritian local radio was established by Charles Jollivet in 1927 which broadcasted music and news for only 2 h daily. Afterwards, in the midst of the Second World War, three radio stations were set up by the British in Mauritius, one amongst which became the national broadcasting station. Following the emergence of radio stations, TV broadcasting became the following invention and the Mauritius Broadcasting Corporation (MBC) was established in 1946 which is as of date the national TV broadcasting station in the country. In fact, liberalisation of the press and media over the years has led to the emergence of some private radio stations and accordingly, there are six private radios in Mauritius namely: Radio One, Radio Plus, Top FM, Planet FM, Wazaa FM and HIT FM. In conjunction, there are six government radios: Radio Maurice, Radio Mauritius, Best FM, Kool FM, NRJ Maurice and Taal FM. As for television channels, there are nine government channels: MBC 1, MBC 2, MBC 3, MBC Digital 4, Sports 11, Ciné 12, Senn Kreol, Bhojpuri Channel and MBC Sat and simultaneously, a collection of channels offered by private televisions such as My.T and Canal+ .

Therefore, to limit the powers of the individual audio and visual media outlets, the Independent Broadcasting Authority (IBA) Act was enacted in the year 2000. Pursuant to the IBA Act, a media regulatory body under the name of the IBA was established to promote, preserve and set standards for media content, impose limitations on media outlets' functions and take such action as it deems fit to ensure impartiality and accuracy. Infringements, as per the IBA, will lead to a fine not exceeding Rs 100,000 and to imprisonment for a term not exceeding 2 years on conviction (Section 37(2) of the IBA Act 2000). All of these measures contributed to Mauritius ranking 61 out of 180 countries by the RSF World Press Freedom Index, as of date. Additionally, to further regulate media content and address the loopholes created by technological evolution, some amendments to the IBA Act were voted on 30 November 2021 by the Parliament of Mauritius, although not without a debate. In fact, these new amendments aroused much controversy amongst the public, who took the view that the measures were out of proportion and unreasonable, with activists such as Bruneau Laurette alleging that there will soon be a social crisis in terms of freedom and social issues (L'Express, 2021).

Therefore, this paper seeks to critically analyse the legal rules pertaining to the media in Mauritius, more specifically to discuss the powers of the Mauritian IBA and the salient features of the new amendments made to the IBA Act. To achieve this research objective, the black letter method will be adopted by analysing relevant legislation and a comparative analysis will be conducted on the powers of the Mauritian IBA against the corresponding powers of the media regulatory body of South Africa. This country has been selected for the comparison because it is imperative to demonstrate how one of the African continent's most economic powerhouse is regulating media as compared to Mauritius. Thereafter, the research will suggest some recommendations for Mauritian stakeholders to adopt in order to foster a conducive environment for encouraging freedom of the media in the country. While the first part of the paper has set out the context of the research including the research objectives and methods, the second part of the paper will discuss some existing literature highlighting the importance of freedom of the media and will discuss the relevance of media regulation. Part 3 will critically analyse the laws of Mauritius on media and the salient features of the amendments made to the IBA Act while Part 4 will examine the laws of South Africa on media regulation. Part 5 will conduct a comparative analysis between Mauritius laws on media regulation and that of South Africa and Part 6 will finally conclude the paper.

At present, there are few literature on the researched topic and this study will be amongst the first academic writings on the effectiveness of the new IBA amendment act. The study is carried out with the aim of combining a large amount of empirical, theoretical and factual information that can be of use to various stakeholders and not only to academics.

3.2 Literature Review

3.2.1 Importance of Media Freedom

In fact, the terminology press implies print media being synonymous with printing press such as newspapers and magazines while media is a much broader term that encompasses broadcasting methods of communication like the TV, radio and internet (Becker et al., 2007). Fundamentally, in today's digitalised and modernised era, it is less likely that people receive news through reading and hence, it is justifiable to admit that the word press will extend to all sources of audio or visual broadcast apart from the written press (Fenwick, 2006). Accordingly, the literature review part of this paper will refer to the word press to refer to both written and audio-visual press.

Indeed, there is a wealth of studies highlighting the importance of press freedom and undoubtedly, free and independent media are the underlying foundation of any democratic state (Organisation for Security & Co-operation in Europe, 2022). This is because freedom of the press fosters good governance principles. For instance, apart from acting as a watchdog to the government for exposing the wrongs committed by the latter while holding it accountable to the public, a press institution also provides a platform focussing on social problems and suggesting ways or even prompting the relevant authorities to take actions to remedy the issues (Sen, 2013).

Consequently, while freedom of the press is considered as crucial for development, Oztunc and Pierre (2021) have set out four distinguishing reasons for advocating in favour of press freedom. Firstly, the authors believe that freedom of speech and public communication is the freedom of human beings which is thus an essential feature for development. Secondly, a free press is much able to disseminate knowledge which is fundamental to policy development formulation through the conduct of public debates which will then lead to a proper diagnosis of the prevailing issues and loopholes in policies. Thirdly, press freedom is purported to give voice to the less well-off ones of the society and fourthly, a free press contributes to a large extent to the generation of new ideas while simultaneously sharing public standards through open public discussions.

Additionally, Kolstad (2011) is of the view that the civil society needs to imperatively be well-informed on governmental actions and the associated consequences. The author thus believes that press freedom has an enormous role to play in promoting a well-educated and well-informed electorate such that the right decisions are made by citizens in electing the country's representatives. Along the same lines, the investigatory skill exercised by the press usually uncovers any autocratic authority and by sharing such information with the public in general, any abuse of power is held accountable by public criticism. This role is particularly vital in the context of electoral campaigns since a free media provides the public with imperative information to compare and evaluate in the light of policies or incentives undertaken or even information on leadership and characteristics of parties/candidates to the election. Therefore, through this function, Norris (2006) qualifies a free media as an agent providing an appropriate platform for a civic forum.

Furthermore, a free press is usually portrayed as an agenda-setter for policymakers (Camaj, 2020; Oztunc & Pierre, 2021). It follows that in addition to channelling citizens' issues to decision-makers, the media also helps in acting as an agent communicator to the government in instances of natural disasters. For example, during the Katrina disaster in the US back in 2005, poor weather conditions had weakened governmental communication networks which made it difficult for people to reach out for help to formal authorities. Accordingly, media reporters were able to act as facilitators to the persons concerned by firstly providing effective emergency relief and secondly, by giving timely and accurate information regarding the evolution of the disaster in question. Admittedly, it can be said that a free media will be able to act as an important channel of information for policymakers thereby prompting governments to be more responsive to people's needs.

Nevertheless, in some instances, the abuse of power has been witnessed by some media operators such as the diffusion of fake news, the promotion of unethical practices, or defamation. Consequently, in several parts of the world, it has become incumbent to regulate the media or press but quite often, it has unfortunately been noted that governments are controlling the press with the view of reinforcing autocratic regimes and to prevent criticism of the government in force (Okeibunor et al., 2021). This is being carried out through censorship, nationalisation of main television and radio channels and even newspaper publishers from the private sector, enactment of stringent laws, or even the use of violence against journalists and broadcasters

(Sussman, 2001). In doing so, some international covenants protecting freedom of expression and speech may be violated. The following section seeks to discuss the relevance of regulating the press from a legal perspective.

3.2.2 The Relevance of Media Regulation

Undoubtedly, the media plays an essential role in investigating, scrutinising and monitoring decision-makers and those in positions of power (Carnegie Trust, 2016). Even international conventions and courts have extended the fundamental right of freedom of expression to the liberty of the media in sharing information. For instance, Article 19 of the Universal Declaration of Human Rights (UDHR) empowers every person to have the right to freedom of opinion and expression and to receive and share information through any media. This particular article is further replicated in Article 19 of the International Covenant on Civil and Political Rights (ICCPR) but with a more elaborated explanation of the methods of communication and some restrictions on the exercise of these rights. In particular, Article 19 of the ICCPR adds to the corresponding article of the UDHR by precising that media communication channels may be either oral, in writing or print, in the form of art or any other media of the person's choice. Moreover, Article 19 of the ICCPR differs from Article 19 of the UDHR by mentioning that the right to freedom of expression of the press shall be exercised by respecting the rights and reputations of others and for the protection of national security, public order or public health and morals.

Furthermore, under Article 10 of the European Convention on Human Rights (ECHR), any person is entitled to hold opinions, receive and disseminate ideas and information without interference. Essentially, "ideas" and "information" referred to in Article 10 of the ECHR not only encompass inoffensive or indifferent content but they also include stuff that may shock, offend or disturb the State or any sector of the population (Muller & Ors v. Switzerland (1988)). Similarly, Article 9 of the African Charter on Human and Peoples' Rights provides for the right to receive information, to express and disseminate opinions within the law. Nevertheless, this legal provision has been criticised by various scholars such as Kas (2021) and Fielden (2012) on the ground that it does not provide the freedom of expression as afforded by other international instruments such as the UDHR, the ICCPR or the ECHR. This is because firstly, there is no corresponding right to impart information and secondly, the right of expression and dissemination has been curtailed with the proviso "within the law".

From the above explanations, it is noted that international law guarantees the freedom of expression by legitimating the right to receive information and ideas. Consequently, when the functioning of the media is curtailed, the rights of various stakeholders to receive information and ideas are being violated including but not limited to the journalists, publishers and even citizens (Kas, 2021). As such, there is a direct link between the expressive and information rights of individuals and the media since the freedom of expression provided under international law also inherently protects the freedom of expression of the media as well. However, there are some instances under which the exercise of the right to express freely may be curbed down at the domestic level, which the following section intends to discuss.

3.3 Media Regulation in Mauritius

3.3.1 Media Freedom in Mauritius

At the domestic level, freedom of the media is protected by Section 12 of the Constitution of Mauritius, which is the supreme law of the country. This particular Section 12 has been inspired by Article 10 of the ECHR and accordingly, the Supreme Court of Mauritius (SCM) relies on judgements rendered by the European Court of Human Rights when considering disputes regarding freedom of expression. It is also apposite to note that the SCM had inferred in the Mauritian cases of Duval v. Police and Hossenbaccus v. Le Mauricien, that the freedom of the media is not higher than freedom of expression afforded to an ordinary individual. Essentially, the SCM had opined in these cases that freedom of expression is an essential instrument for the advancement of knowledge and must be well-guarded to enable the press to fulfil its social obligation to the public.

Generally, any undue interference with the right to freedom of expression is frowned upon. Indeed, pre-publication censorship is thus said to be incompatible with the concept of media freedom (Ramgati, 2015). Nevertheless, the SCM mentioned in the cases of Duval v. Police and Soornack v. Le Mauricien Ltd that since pre-publication censorship is not prohibited by the Constitution, the press can therefore be subject to that restraint. In fact, prohibition on pre-publication is possible if an interlocutory injunction is granted by the court but prior to granting this restrictive order, the court usually takes a reasonable approach by considering if the purported publication will entail serious breaches of a person's privacy or will encourage unethical press practices (Ramgati, 2015). Hence, it can be fairly said that the right to freedom of expression is not an unfettered right and the SCM had summarised in the case of DPP v. Boodhoo that the media freedom is limited by the right to privacy, the law of defamation, the law of contempt of court and the regulatory laws of the media.

3.3.2 Limitation on Media's Freedom in Mauritius

In fact, the right to privacy is a constitutional right guaranteed by Section 9 of the Constitution of Mauritius but this legal provision affords protection against only one type of intrusion in the private life of a person which is restriction against unlawful search on his person or entry on his property without his consent. This particular

Section 9 poses a serious concern for investigative journalism since journalists have to go on-site without permission to uncover ill-doings in society. Moreover, the right to privacy is also provided by Article 22 of the Civil Code of Mauritius, which is reproduced from Article 8 of the ECHR and affords every person to have a private life. While the definition of a private life is not given by the laws, French courts have interpreted it as the right to protection of family life and intimacy of one's personal mails and private conversations and keeping confidential the medical secret of a patient (Lebrun, 2015).

In addition to the right of privacy, the exercise of media's freedom of expression is also curtailed by the laws of defamation in Mauritius. While defamation is the publication or diffusion of false statements without legal justification, these statements are likely to cause prejudice to the victim in society. Accordingly, the author of a defamatory statement can either be criminally prosecuted under Section 288 of the Criminal Code or can be sued under Article 1382 of the Civil Code of Mauritius for a civil action. Nevertheless, a media operator may be exempted from criminal and civil liability in the case of defamation if it is successfully proven that either the information was necessary for public interest and it was based on true facts, or the journalist has acted in good faith and in accordance with the ethics of journalism and the defamatory remarks are based on true facts.

Furthermore, the media's freedom of expression is limited on the grounds of avoiding a contempt of court's offence. In fact, any person who impedes the fair administration of justice will be guilty of this offence and even journalist may fall prey sometimes. For instance, the media was accused in the case of DPP v. Ahnee of making insulting remarks in respect of a judge that had casted some doubts regarding his independence and impartiality. The relevant media operator was thus sanctioned for having scandalising the court. Additionally, in the case of Procureur General v. Delaroche, a media operator was found guilty of a contempt of court's offence for having influenced a pending trial that had interfered with the administration of justice in this particular case. The rationale behind these convictions is to avoid "trial by media" which will negatively affect the right to a fair trial of an accused.

Apart from the above-mentioned legal restrictions on media freedom, the regulatory body of the media also has the power to some extent to curtail the functions and roles of its licensed media operators.

3.3.3 The Regulatory Framework for Media in Mauritius

Basically, almost all countries across the globe have an institutional framework to regulate the behaviour of media enterprises and the main functions of a media regulatory body are to grant licences, provide training to uphold journalistic ethics and standards and to deal with any complaints towards the press (Ramjeet, 2013). Likewise, in Mauritius, it is the Independent Broadcasting Authority (IBA) which is responsible to regulate the audio and visual press that is, radio and television as the sources of information. It is noteworthy to highlight that the written press in Mauritius is unregulated in the sense that there exists neither a voluntary self-regulating body nor a body established by statute to regulate written media, which is not the case for audio-visual media.

In essence, the IBA is established under the IBA Act and in addition to the responsibilities of granting licences and regulating the content of media's diffusion, the IBA is also entitled under Section 4 of the IBA Act to ensure fair competition between broadcasting licencees, set acceptable standards for programmes and advertising, inquire into public complaints against a licencee, promote the efficient use of broadcasting frequency bans amongst others. In exercising its duties, the IBA is empowered under Section 5 of the IBA Act to give written instructions to any licencee and failure by the latter to comply with the directives given, is an offence. It is apposite to note that the IBA has the power to enter the premises of any licencee in order to inspect materials or take possession of the recording of any broadcast programme when exercising its functions and duties (Section 5(3) of IBA Act). Accordingly, it is noteworthy to highlight that as mentioned above, a person has the constitutional right under Section 9 of the Constitution of Mauritius to restrict "*unlawful*" searches on his own property but the particular Section 5(3) of the IBA Act implicitly infer that the IBA has the legal right to enter the premises of its licencees.

Additionally, a new amendment has been made to this Section 5 allowing the IBA to impose administrative penalties on the licencees when exercising its regulatory functions (Section 5(5) of IBA Act). Although the maximum penalty that can be levied is MUR500,000 (USD11,351) as per Section 29(4) of the act, there is no adequate and sufficient information which is available in the public domain regarding the imposition of the quantum of such penalty in order to match the amount with the alleged breaches. In other words, there is no clarity on the type of offences and the determination of the penalty amount. Nonetheless, the principle of proportionality remains applicable since the fine may range between MUR1 (USD0.023) and up to MUR500,000 (USD11,351) (Chan-Meetoo, 2021). Another notable amendment made to the same Section 5 of the IBA Act is the empowering of the IBA to issue and amend the existing code of ethics and code of advertising practice for its licencees without public consultation. In contrast, prior to the new amendments, the IBA had the responsibility to seek the public's views in establishing codes of conduct but at present, the co-regulation between the IBA and the civil society no longer exists which is unfortunate since this is synonymous to autocracy.

Moreover, in connection with its regulatory functions, the new amendments to the IBA Act now permit the IBA with the objective of exercising its powers, to apply to a Judge in Chambers for an order to direct any person to produce records, documents or articles who refused on the ground of confidentiality (Section 18A of IBA Act). However, the ECHR (2021) in its factsheet for the protection of journalistic sources has highlighted that the protection of sources of information of journalists is one of the fundamental baselines for press freedom and it is only in extreme circumstances like for public interest purposes where the disclosure of the underlying source of information may be divulged. To this effect, the ECHR had ruled in the case of Goodwin v. The United Kingdom (1996) that an order requiring a journalist to reveal his source of information concerning a company's confidential corporate plan and the imposition of fine on the latter for having refused to do so was regarded as a violation of his right to freedom of expression under Article 10 of the ECHR. Accordingly, there is hope that Mauritian judges will resist any unjustified request for journalistic sources and will consider the application of this Section 18A of the IBA Act only if the national security of the country is at stake (Chan-Meetoo, 2021). Nonetheless, this provision makes it inconvenient for journalists to conduct investigative work since they are not in a position to secure the anonymity of their sources including whistle-blowers for fear of being identified through raw documents relied upon by journalists.

In addition to the above-mentioned powers, apart from considering applications for licences, the IBA is also empowered to renew existing licences under Section 22 of the IBA Act. While radio broadcasting licences are valid for a one-year period (prior to the new amendments, it was a 3-year period), television broadcasting licences are of a maximum duration of 5 years. This one-year period for a private radio is of too short duration as compared to licence renewal conditions in other countries like the US and France (Conseil Superieur de L'Audiovisuel, 2021). Essentially, in considering an application for renewal, in addition to assessing the applicant's past conduct, the IBA must consider whether the applicant has any pending judicial process against it and if any sanction has been imposed on the latter. While these new amendments will undeniably increase the bureaucracy burden of media operators, it is implied that the government was targeting a specific private radio which regularly diffuses contents against the political party in force (L'Express, 2021) and it is anticipated that the non-renewal of licence is a form of punishment for having illustrated the wrongdoings of the government.

Furthermore, in its capacity as a regulator for the media in Mauritius, the IBA has the power to hold disciplinary proceedings under Section 29A of the IBA Act in the event that a licencee has contravened any provision of the act, has not complied with the laws of Mauritius or the conditions of the licence, has committed a financial crime or is not a fit and proper person. Any subsequent decision by the IBA further to the disciplinary proceedings which may involve the imposition of an administrative penalty shall be published in the Government Gazette and if the licencee feels aggrieved by the IBA's decision, it has the possibility to appeal to the Independent Broadcasting Review Panel (IBRP) within 21 days of the issue of the written notification. In fact, the IBRP is a new insertion to the IBA Act which has the mandate of reviewing the decisions of the IBA. Yet, the independence of this committee is left to be determined.

Primarily, the IBRP is comprised of a chairperson qualified as a barrister with at least 5 years standing at bar and who is appointed by the Minister of Information Technology, Communication and Innovation, and 2 members appointed by the same Minister who have experience in broadcasting and media areas. Moreover, the quorum fixed to hold proceedings of the IBRP is 2 members as per Section 30A(3)(b) of the IBA Act. Arguably, the composition of the IBRP and the quorum may each pose a problem to the effect that the size of this committee appears small and the law is silent as to how the decision will be made in the event of only 2 members present who have dissenting views. Also, no clarity is provided by the law as to whether any other person can act as a proxy to the chairperson in a proceeding of IBRP and whether the proxy also needs to be a barrister with 5 years experience. Furthermore, the IBRP is empowered to hold private hearings at their discretion and no further guidance is provided regarding the circumstances which the review panel considers in conducting proceedings in private and not public. Indeed, the dissemination of facts or circumstances of each case and the judgement provided by the IBRP would have shed light on journalism standards expected by the IBA which will be in the public interest with the view of promoting accountability and transparency.

Undeniably, the new amendments made to the IBA Act have given the media regulatory body in Mauritius huge discretionary powers, the correct application of which remains to be assessed in the coming years. Hence, to investigate where the country stands in terms of media regulation, it is imperative to examine the laws of another country to see if improvements may be made to the media regulatory landscape and the selected country is South Africa which the following section will elaborate on.

3.4 Media Regulation in South Africa

The broadcast regulatory system in South Africa comprises of both a compulsory and voluntary blended approach which makes it an appropriate basis for comparison against Mauritius broadcasting regulatory framework. In fact, a scrutiny of the South African broadcast regulatory system is of utmost relevance in showing the case of the co-existence between state-sanctioned regulation with independent self-regulation.

3.4.1 The BCCSA as the Independent Self-Regulatory Body

Primarily, one amongst the 2 regulatory broadcast bodies is the independent Broadcasting Complaints Commission of South Africa (BCCSA) which concerns those media outlets that voluntarily subscribe to one of its codes of conduct (Milo & Stein, 2022). The basic functions of these codes are generally to set some minimum compliance requirements for the media operators well as to regulate the content diffused by the media. For instance, the Free-To-Air Code (FTA Code) requires those broadcasting licencees who have agreed to fall under the purview of the BCCSA to avoid violence and hate speech (Section 3 of the FTA Code), not to broadcast harmful and disturbing materials to children, prohibit the licencees concerned to diffuse child pornography, bestiality, explicit sexual conduct and violence (Section 8 of FTA Code), and to broadcast news which are true, accurate and fair (Section 11 of the FTA Code). Moreover, the FTA Code imposes an obligation on media operators which choose to be regulated by the BCCSA to exercise considerable care and consideration regarding the privacy, dignity and reputation of individuals while simultaneously acknowledging the fact that these rights may be overridden by a legitimate public interest (Section 15(1) of the FTA Code).

In essence, as mentioned above, subscription to the BCCSA codes is voluntary but in practice, media operators are highly encouraged to subscribe to anyone of them. This is because Section 54(1) of the South African Electronic Communications Act 2005 requires all broadcasting licencees to adhere to the Code of Conduct for Broadcasting Services (which is established by the Complaints and Compliance Committee of the Independent Communications Authority of South Africa (ICASA)) but a leeway is provided under Section 54(3) of the same Act exempting those media operators to comply with the ICASA Code of Conduct if the latter falls under the purview of the BCCSA. Consequently, apart from the FTA Code, the BCCSA has put in place its own Code of Conduct for its broadcasting service licencees. This particular Code provides restrictions on broadcasting some particular content like incest, rape or sexual content (Section 9 of the Code of Conduct), or information advocating war, violence and hatred (Section 10 of the Code of Conduct), mechanisms on the classification of programme as per age or content (Section 14 of the Code of Conduct), the establishment of parental control mechanism (Section 21 of the Code of Conduct) and restriction on public comments (Section 28 of the Code of Conduct), amongst others.

Basically, it is essential to highlight that the BCCSA is an independent body established by National Association of Broadcasters which derives its foundation from Section 34 of the South African Constitution that provides the right of every individual to resolve disputes by an independent and impartial tribunal or forum (BCCSA, 2022a). Consequently, the constitutional document of the BCCSA mentions that the independence of this body is absolute and that it is not accountable to any organisation. Therefore, the BCCSA has established its own complaints resolution mechanism and essentially, any complaint against a licencee has to be in writing and needs to be related to the breach of one of its codes' provisions (BCCSA, 2022b). As a first step, the complaint is evaluated by the registrar who will determine if a prima facie contravention of the relevant code has occurred. In the affirmative, the matter is referred to a commissioner who after considering the facts and written arguments of both the complainant and the broadcaster, provides a written adjudication (BCCSA, 2022c). However, for more serious matters, the registrar will submit the complaint to a tribunal that comprises of 3 or 5 commissioners. These commissioners are appointed by a committee made up of the chairperson of the BCCSA, chairperson of the National Association of Broadcasters and 2 external members of the BCCSA. At the tribunal, the complainant and the media operator will be represented either in person or by their legal representatives (Milo & Stein, 2022).

Nevertheless, if any of the complainant or broadcaster is not satisfied with the decision of the commissioner or the formal tribunal, then the latter has 5 days to appeal against the decision and this appeal has to be lodged with the chairperson of the BCCSA. The case is considered anew by an appeal tribunal comprised of the chairperson of the BCCSA, one public representative commissioner and one broadcaster representative commissioner (Lion-Cachet, 2020). Ultimately, the decision by the appeal tribunal is binding and can only be challenged in a court of law on the grounds of procedural fairness under the Promotion of Administrative Justice Act.

3.4.2 The Complaints and Compliance Committee (CCC) of the ICASA as the State Regulatory Body

Fundamentally, the CCC is a statutory committee established under the ICASA Act and as mentioned earlier, any media operator which does not subscribe to any of the BCCSA codes, will de facto fall under the jurisdiction of the CCC. In fact, apart from matters relating to the BCCSA codes, the CCC has some authoritative and regulatory powers on all broadcasters irrespective of whether they are members of the BCCSA regarding some specific matters such as breaches of licencing conditions, technical aspects of broadcasting and editorial policy, amongst others. This implies that the CCC usually deals with matters that do not directly relate to media content which is mostly left to the BCCSA subject to subscription to one of its codes and accordingly, the ambit of the exercise of the CCC's powers needs to be discussed.

In particular, the South African case of Freedom of Expression Institute v. Chair, Complaints and Compliance Committee 2011 shed light on the jurisdiction of the CCC to hear disputes concerning editorial policy. In this case, some political commentators were not allowed to take part in debates by the head of the South African Broadcasting Corporation (SABC) news and the CCC had to determine whether the editorial policy of SABC was fair and reasonable. However, the CCC thought that it did not have the capacity to rule over this matter since the issue concerned an internal matter of staff conduct. Yet, this reasoning was found to be flawed by the South Gauteng High Court in Johannesburg which ordered that the matter be heard anew by the CCC but with a new committee of commissioners. Ultimately, it was found that the head of SABC denial was biased since the latter was in personal disagreement with the commentators in question and this decision was in violation with the SABC licencing conditions.

In a nutshell, the CCC is the regulatory body mandated by legislation that acts as a watchdog for matters relating to the issuance of licences and their conditions, dealing with frequency spectrum, shielding consumers from unfair practices, poor quality services or inferior products (Parliament of the Republic of South Africa, 2007). Thus, it follows that the CCC has a well-established set of rules concerning the filing of complaints to be heard thereafter in terms of inter alia legal representation, quorum, place of hearings and recording of proceedings. In contrast to the BCCSA, oral hearings are held for each and every matter (Milo & Stein, 2022) and eventually, the CCC submits its recommendations to the ICASA for further actions. In other words, it is the ICASA that will issue the written decision upon suggestions from the CCC which may include the revocation or suspension of licence, remedial and corrective measures like fines or an order to desist from future breaches, amongst others.

Undeniably, the above discussion of the South African broadcast media regulation has highlighted the relevance of a hybrid system of self and statutory regulation which aims at segregating matters of dispute to be heard by differing panels with the aim of promoting independence and transparency. It is against this underlying principle that the following part of the research will provide recommendations for the broadcast media regulatory landscape for Mauritius.

3.5 Comparative Study and Recommendations

While Mauritius has a statutory institution under the name of IBA to regulate broadcast media, the legal framework has unfortunately not attributed powers in the hands of the private sector or the civil society to participate in promoting and maintaining high media standards. In contrast, South Africa has adopted a system of co-regulation by providing discretion to broadcasters to be self-regulated, that is by subscribing to any one of the BCCSA's codes of conduct and then simultaneously for matters other than those stipulated in the codes, these media outlets will be governed by the CCC of the ICASA which is a statutory body.

Indeed, common consensus agrees that effective self-regulation is the best system for promoting high standards in the media as per Section 9(3) of the Declaration of Principles on Freedom of Expression in Africa issued by the African Commission on Human and Peoples' Rights in 2002. This statement is also in line with the Windhoek Declaration for the Development of a Free, Independent and Pluralistic Press issued by the UNESCO in 1991 which encourages the establishment of non-governmental regulations and codes of ethics in order to defend more effectively the profession and ensure its credibility. Accordingly, reference to non-governmental regulations implies self-regulation which shall be impartial, expeditious and cost-effective while promoting high standards in media in accordance with codes of conduct. Moreover, media regulation needs to rely on solid foundations derived from collaboration between the government, private sector as well as the civil society.

Consequently, this research proposes that the regulatory landscape for the broadcast media be reformed in Mauritius to firstly amend the IBA Act to allow for the setting up of a self-regulatory body to which some media outlets may voluntarily subscribe and this institution will be responsible to issue codes of conduct and ethics and regulate matters regarding the content being broadcasted. In addition, this self-regulatory body will be instituting its own disciplinary and dispute resolution mechanism and this framework will in turn provide the IBA with adequate time to monitor media operators on their licencing requirements, technical aspects of broadcasting and editorial policy amongst others. As such, the broadcasters and the civil society will each have separate adjudication bodies to hear disputes and this will enhance transparency and increase independence. Moreover, it is also recommended that the public in general be consulted prior to finalising codes of conduct which is not the case for the codes being issued by the Mauritian IBA despite the fact that media regulation is a matter of each and every one of us since a public service is concerned.

In fact, this approach may be regarded as a multi-stakeholder regulation which is complementary to co-regulation since the input of the government, private media outlets as well as the civil society, will be considered.

In addition to the media regulatory landscape, it is apposite to compare the composition of the complaints committee of the Mauritian IBA against the South African BCCSA as well as the CCC each to suggest recommendations with the view of improving independence of the adjudicating committees. While under the IBA Act, the IBRP is comprised of 3 members all of whom are appointed by the Minister of Information Technology, Communication and Innovation, in South Africa, the commissioners of the BCCSA, are representatives of the BCCSA itself, the National Association of Broadcasters and 2 external members of the BCCSA. Similarly, the CCC members are appointed by the ICASA upon fulfilling some prescribed conditions under the ICASA Act. Consequently, it is crystal clear from the composition of the Mauritian IBRP that the government wishes to hold a controlling power on the decision-making process of this committee and this in turn questions the independence of the IBRP. It is imperative to highlight that existing literature has previously concluded that autocracy in complaints committee often leads to the penalisation of a media operator who was critical to the government or the political social system (Lion-Cachet, 2020). Hence, it is hereby suggested to review the legal framework governing the appointment of IBRP members to allow for members of civil society or even the opposition party to take part in selecting these members. This approach will ultimately result in a more transparent and independent system.

Lastly, regarding the media's diffusion or publication liberty, it is incumbent to assess the relevance of the limitations of freedom of expression in the current context as of date. In other words, what was considered to be irrelevant by societal norms previously may now be of huge importance in today's society. For instance, much of what is now accepted would have shocked the community to the core a generation ago. One example is the publication of the Lady Chatterley's Lover by Penguin Books in 1960. This book narrated the story of an extra-marital affair between Lady Chatterley and her gamekeeper, and comprised of explicit descriptions of sex including reference to annal sex which was illegal at that time in the UK. The book was published in Europe in 1928 but not in the UK until the year 1960. Immediately after its publication, a case was entered against the publisher Penguins Book under the Obscene Publication Act 1959. Under this Act, an article was deemed to be obscene if it has an effect of tending to deprave and corrupt persons who will read, see or hear the matter contained in the same article. Therefore, jury members had to decide whether the Lady Chatterley's Lover book satisfied the definition of obscene and then, whether the publication was justified as being for the public good on the ground that it is in the interests of science, literature, art or learning or other objects of general concern. Ultimately, jury members returned a verdict of not guilty and this case consequently created a landmark decision to the effect that if a work is of literary merit and for public good, people are free to publish and diffuse such material. Moreover, the case illustrated the changes in society in tolerating the representation of sex and sex relations in publication since human beings are constantly evolving across cultures which have left behind taboos that were once inherent in the societal system.

Similarly, it becomes vital to consider the tolerance level of the existing modern society in which we are living to evaluate whether it is relevant to ease the freedom of expression against the threat of censorship by laws and regulations.

3.6 Conclusion

Undeniably, press freedom is accompanied by numerous responsibilities. For instance, the rights of others must not be infringed when the media is exercising its constitutional right of expression. Likewise, since the media holds the power of scrutiny over others, it should primarily and essentially be subject to evaluation and monitoring itself. This consequently explains the importance of media regulation but having an autocratic legal and regulatory framework may be detrimental to media freedom in any country. In the context of Mauritius, the IBA Act was under the dire need of amendments to align with new trends in media operations which explains why the Mauritius legislator has proceeded with approving the IBA legislative amendments despite contentions from numerous parties. Nevertheless, there are several issues associated with the majority of amendments which have been critically assessed in this paper in terms of inter alia the administrative penalty, request for journalistic sources of information, shortening of licence renewal period, disciplinary proceedings to be heard in private and the composition of the IBRP. As such, this critical analysis forms the underlying basis further to which some of these legal provisions need to be rectified in order to promote a more free, democratic and independent media regulatory landscape. In this regard, following the comparative study conducted in this paper, this research suggests that a multi-stakeholder approach be adopted to consider both self and statutory regulation over media operators in Mauritius with different avenues for hearing disputes likewise the case of South Africa. In addition, the composition of the IBRP needs to be modified to allow for some members to be appointed by external parties such as the opposition party, the Chief Justice or even the Director of Public Prosecutions instead of only the government nominating these members.

References

Becker, L., Vlad, T. & Nusser, N. (2007). An evaluation of press freedom indicators. *International Communication Gazette, 69*(1). https://doi.org/10.1177/1748048507072774 [Accessed on 24 January 2022]

BCCSA. (2022a). *Constitution.* https://bccsa.co.za/bccsa-constitution/ s1 of preamble [Accessed on 14 February 2022]

BCCSA. (2022b). *Criteria for a complaint.* https://www.bccsa.co.za/criteria-for-a-complaint/ [Accessed on 14 February 2022]

BCCSA. (2022c). *Adjudication process.* https://www.bccsa.co.za/frequently-asked-questions/ [Accessed on 14 February 2022]

Camaj, L. (2020). The media's role in fighting corruption: Media effects on governmental accountability. *The International Journal of Press/Politics, 18*(1). https://journals.sagepub.com/doi/abs/10.1177/1940161212462741 [Accessed on 20 January 2022]

Carnegie Trust. (2016). *Regulation of the Press.* https://d1ssu070pg2v9i.cloudfront.net/pex/carnegie_uk_trust/2016/02/pub1455011678.pdf [Accessed on 20 January 2022]

Chan-meetoo, C. (2021). Assessing the Independent Broadcasting Authority (IBA) Amendment Bill 2021. *Le Mauricien.* https://www.lemauricien.com/actualites/assessing-the-independent-broadcasting-authority-iba-amendment-bill-2021/461208/ [Accessed on 7 December 2021]

Conseil Superieur de L'audiovisuel. (2021). *Autorisation des radios publiques.* https://www.csa.fr/Reguler/Creation-et-regulation-d-une-radio/Les-appels-a-candidatures-pour-les-radios-FM [Accessed on 31 January 2022]

DPP v. Ahnee [1992] MR 184

DPP v. Boodhoo [1992] SCJ 472

Duval v. Police [1974] MR 130

ECHR. (2021). Protection of journalistic sources https://www.echr.coe.int/Documents/FS_Journalistic_sources_ENG.pdf [Accessed on 31 January 2022]

Fenwick, H. M. (2006). *Media freedom under the human rights act*. Oxford University Press.

Fielden, L. (2012). *Regulating the press: A comparative study of international press councils.* Reuters Institute for the Study of Journalism. https://d1wqtxts1xzle7.cloudfront.net/33887914/Regulating_the_Press-with-cover-page-v2.pdf?Expires=1643092924&Signature=HrVhHtcObsrUxs-fB7UZGXwBTbjHU15FDMYZUJsg78S19eTEk7Qcfl-lznt6HyUtRjIaz9633Cvb0ui794QV34n0RhE3GoVMT9RpksG8W7a3oRTsTScJePC49jGAyLmlIB5J5nwjgAwUew2rYp-58TlmvZGTD6GQgDEdH9Y8rtUwE98ML7Nv3fdvqjBEo7kTX~y2gtaRvIwErGfOMtYSYPpSv2JokGziu-q-qe6pEABvMj5MmCQpBbzCJK5lD6AFA4k5wXt-OdNX45sfWLV-FwWPOntGW2IaVIlE~3X4TH092HAEDCD-dOYYUaMUY3oYu0LLX9PMUoAfqZ9-rO6QQ__&Key-Pair-Id=APKAJLOHF5GGSLRBV4ZA [Accessed on 25 January 2022]

Freedom of Expression Institute v. Chair, Complaints and Compliance Committee [2011] JOL 26704 (GSJ)

Goodwin v. The United Kingdom [2016] ECHR Grand Chamber 17488/90

Hossenbaccus v. Le Mauricien [2001] SCJ 283

Kas, A. (2021). The role of the media and press freedom in society. In *Media law handbook for South Africa,* Konrad Adenauer Stiftung.

Kolstad, I. (2011). *Does democracy reduce corruption?* Working Paper, Chr. Michelsen Institute. WP 2011:4. https://www.cmi.no/publications/file/4315-does-democracy-reduce-corruption.pdf [Accessed on 20 January 2022]

L'Express. (2021). *Independent Broadcasting Authority Act: A law that is causing a lot of ink to flow.* https://www.lexpress.mu/article/402266/independent-broadcasting-authority-act-une-loi-qui-fait-couler-beaucoup-dencre [Accessed on 7 December 2021]

Lebrun, P. (2015). La vie privée. In *Travail social: le moment de transmettre.* In Empan. France: Eres

Lion-Cachet, F. (2020). *An analysis of press regulation and the proposed media appeals tribunal in line with constitutional imperative of a free and independent press.* Level II Dissertation Submitted in Fulfilment of the requirements of LLM in Perspectives on Law. North-West University, South Africa

Milo, A., & Stein, P. (2022).*A practical guide to media law.* https://www.bccsa.co.za/frequently-asked-questions/ [Accessed on 14 February 2022].

Muller and Ors v. Switzerland (1988). *European Court of Human Rights,* Case No 25/1986/123/174

Norris, P. (2006). *The role of the free press in promoting democratization, good governance and human development.* Working Paper, Human Rights Initiative, Harvard University, WP. 2006/13. https://www.humanrightsinitiative.org/programs/ai/rti/articles/undp_rti_2006/annex3_background_paper.pdf [Accessed on 20 January 2022]

Okeibunor, A., Kamal, P. & Ally. A. (2021). Media and protection of human rights. *Ianna Journal of Interdisciplinary Studies, 3*(2), 115–128. https://iannajournalofinterdisciplinarystudies.com/index.php/1/article/view/62 [Accessed on 20 January 2022]

Organisation for Security and Co-operation in Europe. (2022). Media freedom and development. https://www.osce.org/media-freedom-and-development [Accessed on 20 January 2022]

Oztunc, M., & Pierre, M. (2021). Analysis of the obstacles to the freedom and independence of the media in the world and Turkey. *The Turkish Online Journal of Educational Technology, 20*(1). https://files.eric.ed.gov/fulltext/EJ1290795.pdf [Accessed on 20 January 2022]

Parliament of the Republic of South Africa. (2007). *Report of the ad hoc Committee on the Review of Chapter 9 and Associated Institutions.* A report to the national assembly of the parliament of South Africa. Cape Town, South Africa.

Procureur General v. Delaroche [1893] MR 13

Ramgati, P. (2015). *Freedom of the media in Mauritius.* Level III Dissertation submitted the partial fulfilment of LLB. University of Mauritius, Mauritius.

Ramjeet, M. A. (2013). The freedom of expression with regards to the media: A fundamental element of democracy. *International Journal of Law, Humanities and Social Science, 6*(3).

Ruckenstein, M., & Turunen, L. M. (2019). Re-humanising the platform: Content moderators and the logic of care. *New Media & Society, 22*(6). https://doi.org/10.1177/1461444819875990 [Accessed on 20 January 2022]

Sen. P. (2013). Press freedom: What is it good for? *Index on Censorship, 42*(3). https://doi.org/10.1177/0306422013503105 [Accessed on 20 January 2022]

Sindane, S. (2018). Press regulation in South Africa and its implications for press freedom. *Communitas, 23*(2018). https://doi.org/10.18820/24150525/Comm.v23.10 [Accessed on 20 January 2022]

Soornack v. Le Mauricien Ltd 2013 SCJ 58

Sussman, L. R. (2001). *Press freedom in our genes: A human need.* World Press Freedom Committee.

Chapter 4
Greenstraw: A Social Enterprise Practicing and Promoting Sustainability

Srushti Govilkar, Kritika Jaiswal, Ruchi Tewari, and Aatman Shukla

Abstract Sustainability is a widely popular concept but a complex concept and phenomenon which brings along challenges of implementation. Despite multiple programmes, and discourses available on practicing sustainability, a large number of individuals and institutions continue to be apprehensive about inculcating it as a part of their everyday lives or regular operations, respectively. A beam of light in which situation is the social enterprises, which attempt at bridging the gap between the Not-for-profits (NFPOs) and mainstream profit-driven businesses (Kerlin in VOLUNTAS: Int J Volunt Nonprofit Organ 21(2):162–179, 2010).

The current study attempts to showcase how sustainability (environmental and social) is practised in a social enterprise—Greenstraw is the case in point. Greenstraw has a lifestyle store, cafe, community and coworking space, and a foundation. It practises and promotes sustainability and inclusivity in its business operations. It has managed to offer itself as a successful example of a sustainability-driven social enterprise. Its operations present a perfect mix of competitive product offerings all of which are sourced, produced, and delivered in a sustainable manner. The study attempts to apply Petit's Behavioural Drivers Model (2019) on Greenstraw to understand its impact and what needs to be done in the future.

Keywords Sustainability · Inclusivity · Social enterprise · Behavioural Drivers Model · India · Case study

S. Govilkar (✉) · R. Tewari
MICA—The School of Ideas, Ahmedabad, India

K. Jaiswal
Greenstraw, Ahmedabad, India

A. Shukla
Kochar and Co, New Delhi, India

D. Crowther and S. Seifi (eds.), *Preparing for a Sustainable Future*,
Approaches to Global Sustainability, Markets, and Governance,
https://doi.org/10.1007/978-981-99-2456-1_4

4.1 Introduction

Sustainability is a buzzword, and its implementation has seen various forms. Individuals, institutions, regulatory bodies, and not-for-profits (NFPs) engage in sustainability. It is rooted in the idea of meeting, the needs of the present without compromising the access to it for the future generations for their own needs (Brundtland, 1987). It is the long-term maintenance of responsibility, including taking cognisance of environmental, economic, and social dimensions (Sharma et al., 2014). While the concept of sustainability is broad due to various contexts making it difficult to define, it is certainly not abstract. It will always be defined within the context of its orientation to nature, alternate resources, economic growth, population growth, technology and its role, social equity, etc. (Vos, 2007).

Apart from the environment, sustainability also involves social sustainability which includes heritage and culture. UNESCO World Heritage Convention claims the long absence of heritage from the mainstream debate of sustainable development, even though it is essential to the society and its contribution to the social, economic, and environmental goals. Resultantly, the UN General Assembly integrated the role of culture via cultural heritage and creativity as a facilitator of sustainable development in its 2030 Agenda. Based on the platform provided by World Heritage, a platform could be developed to test newer approaches demonstrating the heritage relevance for sustainable development. SUMAS Team (2021) shares that an essential element of sustainability is inclusion.

Indian social enterprises have always been a layered construct for the practice preceded theories and concepts around it (Shukla, 2011). Apart from that, fusing inclusivity with sustainability has been a challenge for the socio-cultural context of the country is multi-layered in terms of diversity, going beyond the differences of race, gender, age, religion, socio-economic, etc., but including the issues of social context of the country in the form of castes and region (Chakraborty, 1987).

The focus of this study is to look at sustainability and inclusivity as a practice within an Indian social enterprise as a case in point. The social enterprise—Greenstraw has a cafe, a lifestyle store, a community and coworking space, and a foundation. The study analyses its practice of sustainability and inclusivity from the perspective of Petit's 2019 conceptual framework of Behavioural Drivers Model.

4.2 Literature Review

4.2.1 Sustainability

The United Nations Brundtland Commission defines sustainability as "meeting the needs of the present without compromising the ability of future generations to meet their own needs" (Brundtland, 1987). However, it is complex and subjective, given the multi-dimensional nature of it, both in theory and practice. The key dimensions of sustainability are the governance structures, business models, management measurement, and reporting systems and for a practitioner, it becomes challenging to manage a balance between these dimensions given that these dimensions have an intersecting relationship (Giovannoni & Fabietti, 2013; Wilkinson et al., 2001). The key challenge of implementation is due to the limited and vague definition of sustainability (Dixon & Fallon, 2008). This has resulted in multiple similar and allied business practices like social responsibility, environmental management, or business sustainability, all of which have overlaps but are unique in certain ways leading to almost "everything and anything" being categorised as sustainability (Werbach, 2011). This ambiguity has created implementation challenges for firms and various not-for-profits, and society, where both national and international regulatory boards raise questions about the sustainable actions taken by the firms. Responding to the pressures, firms have begun to operate within the context of social and environmental responsibility and focus upon the measurement of sustainable action. Within the Indian context, environmental sustainability has been a part of its history as a practice (Manjrekar, 2017).

4.2.2 Inclusivity

Sustainability also includes social aspects apart from the widely known environmental one. Inclusivity is an integral part of social sustainability with the United Nations 17 Sustainable Development Goals (SDGs) featuring it as a part of its 2030 Agenda. It defines inclusivity as a principle where *every person should reap the benefits of prosperity and enjoy minimum standards of well-being.* Along the similar lines, World Bank (2021) defines social inclusion as *"The process of improving the terms for individuals and groups to take part in society"* and *"the process of improving the ability, opportunity, and dignity of people, disadvantaged on the basis of their identity, to take part in society"*. United Nations defines social inclusion as a ***process of improving terms of participation in society for people who are disadvantaged on the basis of age, sex, disability, race, ethnicity, origin, or economic or other status, through enhanced opportunities, access to resources, voice and respect for rights*** (2016). To avoid the risk of being abstract, the Commission of the European Communities (2003) gave a comparatively sharper definition of social inclusion—as a process ensuring those who risk poverty and social exclusion get the opportunity and necessary resources to fully participate in economic, social, political, and cultural life which enables them to enjoy a standard of living which is normalised in the society they live in. This ensures greater participation in decision-making which impacts their life and access to the fundamental rights. In the Indian context, Prof. T. K. Oommen in one of his special lectures held at the University of Hyderabad cited that many historians argued that the social exclusion in India dates back to the Aryan invasion, followed by the Muslim era, and the colonisation. The series of invasions impacted the social formation determining the issue of insiders versus outsiders and adding layers of social exclusion in the Indian context. Another aspect of social exclusion apart from the invasions was seen in rural India in the hierarchy of the castes, where the upper castes were dominant and other backward classes, untouchables had no say (Rajora & Yallo, 2021).

4.2.3 Sustainability in India—Practice, History, and the Ethos of It

Social responsibility and sustainability are intrinsic in the Indian context. In India, the traditions and culture play an important role and it is extremely critical to inculcate these values and culture in a child at a young age along with the family practising it regularly to strengthen it. There are multiple practices in the culture, which have been a part of daily life and have been practised over the years. The country is extremely resourceful and wastefulness is often viewed as a lack of consideration of others, selfish, and short sighted without keeping the future in mind (Pandey & Sahay, 2022). This gives us a glimpse of the ethos that India has towards sustainability and adds the historical context to it. In terms of practice in domestic life, the use of stainless steel, *kansa* or brass utensils, clay pots, etc., are regularly used, last for generations, and do not add any to any waste unlike plastics. When it comes to mass gatherings where food is served, banana leaves, *sal* leaves stitched together to make a plate called *pattravali* (leaves woven together as a plate) were used in the olden days. Today some of these are making a comeback along with clay *kulhars* (earthen cups), bamboo plates and cutlery, jute bags, cardboard container boxes, etc., due to awareness of the harmful effects of one-time use plastics in the country, and its subsequent ban (Manjrekar, 2017).

However, in ancient India sustainability was not a practice limited to food, but all the aspects of life. In beauty and grooming, wooden or neem combs, milk and turmeric for skincare, soap nuts, and *amla* or Indian gooseberries were some of the ingredients which were regularly used on a regular basis. In terms of clothing, clothes made of natural fibres were widely used. In terms of architecture, many historical monuments reveal sustainability as the rationale behind the creation of the structures. Given the different kinds of terrain and environment that various parts of the country reflect, there is a certain degree of unique regionality reflected in these structures which are suitable to the environment of the region. For example, the Ajanta caves which have vaulted ceilings have sun windows to naturally light up the prayer halls, along with this, the ceilings are low to naturally lower the temperature in the cave. In terms of natural building material, Cob, a mixture of water, subsoil, and fibrous materials like straw, was a popular choice. The construction using the material provides good thermal mass and makes it an insulated warm structure when it comes to colder regions and also earthquake resistant to a certain extent. Some of these structures can be seen in Spiti. In contrast to these structures which keep the interiors warm, *Badgirs*, an ancient technique ensured that the interior of the structure or the building is cool, which became a popular choice of technique to be implemented in hot and dry areas. These cooling structures are seen in some parts of Rajasthan (Dey, 2021). All these examples of practices reflect that in order to follow a sustainable lifestyle you need not compromise on your quality or standard of living (Pradeep, 2020).

However, despite having a rich history and tradition of practicing sustainability, one might wonder what must have surpassed for India to face the challenge of sustainability today. Mehta (2021) argues that multiple invasions followed by colonisation left the country infested with not just different kinds of pollution but with the implications of poverty, and wastefulness.

Pandey (2017) in her article shares that India is one of the least wasteful countries. Owing to multiple developmental schemes which take SDGs as an essential, the country has reflected leadership in issues of climate change and achieving the SDGs, which was also acknowledged at the UN's High-Level Political Forum of 2017.

4.2.4 Social Inclusion in India

The social fabric of India is very diverse and therefore highly complex. While the Western society is divided into the haves and have-nots on the basis of economic well-being, and has some inequities vis-a-vis gender, race, religion, sex, and ethnicity. In India all these social attributes get exacerbated because the inequities are very high. Traditional cultural values clash with the rapidly changing social profile due to globalisation and penetration of digital technology. The caste system adds another layer to diversity. Vulnerable or the marginalised sections of the society include population from Scheduled castes, Scheduled Tribes, other backward castes, etc. Discrimination against the marginalised is quite deep-rooted in the country owing to the unequal social structures and cultural mindset of the masses. Social inclusion in India requires public policies to be effectively implemented, where the major challenge lies in. The Government of India has acknowledged this and in order to curb the issue, has also adopted various policy strategies like Tribal Sub-Plan (TSP), Special Component Plan (SCP), for Scheduled Tribes and Scheduled Castes, respectively. While discrimination at a socio-cultural level is complex to tackle, one of the effective ways to deal with it is via economic policies which aim at equity in terms of access to resources (Shukla, 2020).

4.2.5 Social Enterprise

Social enterprise is seen as a mid-way between a pure-not-profit which is often found to be struggling for funding and often to myopic about the development issues which they champion. In contrast the profit-driven mainstream corporate firms are known to be in an overdrive to increase their profits. While the latter are focused on complete sale of the output in the market, the not-for-profit condemns the profit orientation and focuses on the issues of advocacy and welfare at large. The approach of the corporate results in unethical business practices due to lop-sided benefit to the shareholders, often at the cost of other stakeholders especially the environment (Kerlin, 2010). Social enterprises (SE) often bridge various components and entities of the third sector. SE attempts to bring in the sale of output with limited profit distribution and at the same time attempting to give back to the society including advocating for various issues and causes which aim at the welfare of the society (Kerlin, 2010). Sengupta and Sahay (2017) argue that social enterprises are a hybrid organisation which attempts to achieve both the market orientation and the social value creation for a double bottom line, strongly influenced by contextual, structural, and institutional forces of the region and its perspective. This makes it extremely challenging to define the term social entrepreneurship. Various definitions of social enterprise make the concept quite complex, due to the contextual differences worldwide. EMES, a European research network, has created some criteria for an entity to be recognised as a social enterprise. These criteria are—(a) must be involved in a regular continuous production or service activity to people, (b) must be created by a group of people who have autonomy in terms of projects, (c) there is a significant level of economic risk involved, (d) there is some involvement in paid work, (e) aim to benefit the community, (f) initiatives are launched by a group of citizens, (g) decision-making power is not solely based on the capital ownership, (h) has a participatory approach involving all the stakeholders (customers, users, etc.), (i) limited distribution of profits which is not with the goal of profit maximisation (2022). Furthermore, there are challenges of diverse social entrepreneurial efforts in terms of their approaches to achieve social goals, contextual embedding of the practice, and diversity of the discipline. However, some attempts have been made to categorise the definitions of social entrepreneurship into three categories—social mission, innovative and sustainable solutions, and entrepreneurial strategies (Shukla, 2020).

4.2.6 Social Enterprise in India

India boasts of several social enterprises like Amul, Lijjat, Aravind Eye Care, etc., and Shukla (2020) attributes the emergence and actions of these enterprises to the country's rich culture and tradition which is focused on giving back to the society. Every member of the society is involved in some way to contribute to the challenges faced and for collective well-being (Chakraborty, 1987; McClelland, 1975; Shukla, 2011). These were reinforced further during the country's struggle for independence, where the Gandhian philosophy was to empower grassroots society or the *gram swaraj*. The philosophy was widely promoted and accepted in the country with many custodians carrying out the implementation in the form of *Bhoodaan (donation of land to the landless)* and *Graamdaan (donation of villages to the homeless)*. These continued to be followed even post-independence under the leadership of Prime Minister Pt. Jawaharlal Nehru (Pandey & Sahay, 2022). Moreover, with theories on social entrepreneurship complex in nature, its setting in the Indian context adds a layer of complexity given the country's diverse culture, history, traditions, and practice preceding the theory (Shukla, 2020).

The United Nations World Youth Report 2020 reveals findings from various studies that today's youth are highly motivated to make a positive social change, and that social entrepreneurship is one of the most potent ways of mobilising youth and facilitating their efforts towards achieving major social objectives like employee creation, poverty alleviation, integration, and inclusion, which ultimately help in achieving multiple SDGs. In a young and developing country like India, this becomes crucial.

4.2.7 Sustainability in Social Enterprise

Jayawardhana et al. (2022) in their study argue that sustainability is the core of existence for the social enterprises within the context of social mission. They further highlight the lack of a systematic categorisation of literature for social enterprises and how that poses a challenge for researchers on social entrepreneurship. However, research in the domain of sustainability in social enterprises is in an infancy stage. Kummitha (2017) in the book *Social Entrepreneurship and Social Inclusion—Processes, Practices, and Prospects* highlight the community contribution and how it is embedded in influencing the practice of social enterprises. He argues that the developmental discourse has ignored the community participation in the planning and execution of the projects taken up by the social enterprises for the community's benefit. Resultantly, the efforts to facilitate the economy in developing countries have failed for apart from the community participation, the social enterprises failed to take culture and values into consideration.

4.3 Theoretical Framework

4.3.1 Behavioural Drivers Model

Moving to sustainable action requires a change in mindset, attitude, and behaviour. Ideologically, most people agree with sustainable ways but find it challenging to accept and include it as a way of life. Therefore, the current study used Social and Behaviour Change (SBC) theories. SBC helps to create and conduct participatory situational assessments, design, operationalise strategies, and monitor its extent. It also aims to promote evidence-based programmes guaranteeing the highest standards of attaining sustainability and impact in protecting and promoting children's and women's rights and offers a conceptual framework model in the form of Behavioural Drivers Model (BDM). BDM represents the relationship between various forces that have the potential to affect behaviours within specific contexts. It is voluntary and agnostic in nature and raises the fundamental question of the rationale behind people doing what they are doing. Petit (2019) augmented the Behavioural Drivers Model by further borrowing from theories in the domain of sociology, psychology, and environment. The rationale behind using this model is that it offers a practical approach in terms of assessing the impact by analysing the behavioural responses of the people, and how to incite a behavioural change to bring about a larger social change.

4.3.2 Methodology

This study presents a case study from India about a social enterprise, founded by two young entrepreneurs who were passionate about sustainability and wanted to positively impact the social and environmental ways of the people of their city. A qualitative research method was used to gather data. Semi-structured interviews were conducted with all the three partners. The interviews lasted for an average of 130 min. These interviews were conducted after ethnographic observations were made. The researcher spent 30 min on every alternate day for a period of three weeks. The days were selected in such a way that over a period of three weeks, all the days were covered to ensure that certain customers who visited the cafe or store on any specific day were also covered. The time of the visit was also randomised to cover morning, noon, and evening visits. Notes were made about the customer's demographics, and interaction with the staff and people they came with, and data about the kind of orders or inquiries was taken. On the basis of the observational data, and the BDM model, the questions for the interviews were framed and posed to the founders of the enterprise were framed. A written consent of the founders was taken before the start of the research. All semi-structured interviews were transcribed and then a thematic analysis was made. Sustainability themes were drawn out to understand the match between the vision of the founders and their activities which they discussed in the interviews.

4.3.3 Background

For a country like India whose population is nearly 1.38 billion (Das, 2022) and the responsibility of providing for communities that do not have proper means of health, education, and shelter also fall upon the organisation like social enterprise, NGOs, or for that matter any enterprise who uses the resources of the society to meet their particular ends. When it comes to the corporates engaging in the activities towards society it is often done under the Corporate Social Responsibility (Dahlsrud, 2008).

In an ideal world, the resources allocated should have been equal or at least based on the equity. Hence, the benefits arising from it would have direct nexus to the resources utilised or misused, and the implications or the ramifications of those actions could have been easily traced, and administered with a solution which helps to facilitate optimum utilisation of the resources. Resultantly, the society would have been just and fair and the role of organisations like social enterprises would have been negligible or limited, for then the very nature of any organisation would have the idea of a just world embedded in its foundation. Whereas, the world which we foster in the point of origin for each individual is different, for most of the population in India highly unfair in terms of resource utilisation and the burden to hold anyone/any organisation/enterprise responsible becomes an extremely difficult and cumbersome process. This often increases the disparity between the haves and have-nots where the conflict often arises for the basic necessities to sustain life. Resultantly, while a few abuse the resource and then there are a larger part of the population which is disadvantaged at large. This is the key area where social enterprises become an important tool of change. Their orientation often bridges the gap between making a difference and in turn bringing in a change, and at the same time not essentially compromising on its aim of breaking even. Based on this ethos, lies the entire core of Greenstraw, which it attempts to implement based on its different verticals in order to cater to the larger audience.

Greenstraw is a lifestyle store, cafe, community and coworking space, and green consultant. It is an initiative to create a community of conscious and conscientious individuals by curating low waste and interactive solutions to everyday things to bring about a sustainable behavioural change. The lifestyle store exhibits and sells curated designer wear within the sustainability domain and from artisans across the country working with natural and recyclable textiles and accessories. The cafe menu offers homemade confectioneries, fresh food items from home chefs and bakers. The community and coworking space for freelancers and the same space can also be rented for events that promote the concept of sustainability.

In terms of operationalising, for Greenstraw means to achieve the particular objective is as important as ends to meet along with the steps and strategies to initiate the conversation and take actions around sustainability keeping in mind the UN SDGs, as well as to provide sustainable solutions to everyday living. For any social enterprise to have a long-term existence that too with the least interference or control from other stakeholders of the society, it is important that it is self-sustained. Thus the profit becomes one of the ancillary objective in the long run, primary being to bring about shift in the collective consciousness of the society through conversations and actions and tweaking these two from time to time to meet the need of the hour. For e.g. the distribution of three ply surgical mask during the peak of covid became a necessity which had long-term environmental impact, often the need of the times would dictate the call of action for any social enterprise.

Basically social enterprise could be a business model whose long-term goal is not to profit maximisation or wealth maximisation of its immediate stakeholders but to channelise the resources of the society for the maximum benefit of the society as a whole including environment or giving least amount of stress to environment.

In case of Greenstraw, the profit generated are through providing sustainable solutions to everyday living, which in the present case for now is the cafe, coworking, conscious lifestyle products, and profits generated from these are then utilised for other welfare activities which could be multiple including educating people about sustainability through various drives, plantation drives, enhancing local communities by providing for better healthcare, sanitation, education, and clean environment to foster. Thus making means as important ends.

Essentially, the prime objective of Greenstraw is to practice sustainability and create awareness. At the core, the founders' motivation for starting Greenstraw was to create a community of like-minded people who believed in the notion of sustainability and could come together to have relevant conservations and find power in the ideology. Pooja, one of the founding members smiling, said that, "while we hear so much about sustainability in the media, an individual still feels isolated and at times mocked and foolish if he wants to practise sustainability in his daily life. We wanted to pull in such people to help them believe that sustainability can be practised and go ahead as an enterprise as well". A lot that Greenstraw does in terms of its economic and social activities is a reflection of the founders' personal experiences and a shared passion around sustainability and inclusivity. Jay, one of the founders, strongly believes in the idea of going back to the roots without compromising on the contemporary lifestyle.

In times of difficult situations, Greenstraw chooses to practice sustainability by looking for alternatives to the mainstream non-sustainable products or services. For example, when Greenstraw was started in 2020 (soon after which COVID struck) as an immediate measure, it distributed three ply masks, and soon it switched to other sustainable options like cloth masks which are made out of natural fabrics like cotton, and from industrial scraps of fabric to foster and practice of zero waste. Owing to the agility of the enterprise to offer sustainable solutions even in difficult situations, it has seen a rise in the number of people who visit them. During COVID, after the lockdown was lifted the coworking space saw traction which generated business for the cafe and gradually the lifestyle store also began to see a rise in the footfall. The numbers have grown by 100%.

4.4 Ideology of Greenstraw

The ethos started with the aim of making sustainable living through products accessible to everyone. In order to achieve this as a long-term practice, one needs to influence it as a part of the behaviour. Once it is imbibed in the behaviour, and cemented with frequent interactions, discussions, and media reflecting the same, it becomes a part of the everyday practice (Bandura & Walters, 1977). With this principle, Greenstraw created a space where people can come to holistically understand sustainability.

The experiences and spaces that Greenstraw creates act as a facilitator to bring about a change in the way people act on an everyday basis and create positive action and conversation around sustainability. Each initiative is developed focusing on the intent to bring about systematic social and behavioural change.

4.5 The Greenstraw Foundation

The foundation aims to work towards connecting sustainability and social causes by engaging with individuals and organisations, to achieve seven sustainable development goals—good health and well-being, quality education, sustainable cities and communities, climate action, reduce inequalities, responsible consumption and production, gender equality, decent work, and economic growth. Over the years the foundation has launched multiple interactive community initiatives creating social and environmental impact.

4.5.1 The Green Straw Spaces—Cafes, Stores, and Coworking and Community Space

The spaces are conscious low waste spaces where the purpose is to consume everything in the most sustainable way with the aim that everything goes back to Mother Earth and to give back more than what was consumed. Operations in these spaces are embedded in the regular practices of recycle, upcycle, and compost. The spaces are created in a way that everything around has been thoughtfully curated to create a first of its kind green space in the city. The founders have attempted to create a space and consciously avoid defining it to avoid colouring people's association with it. Hence, the space is a versatile amalgamation of a coworking space, a cafe, an eco-friendly store, and a space to expand and build the community of conscious and conscientious humans who try their best to do more everyday, do something new, to create a sustainable world for all of us.

4.5.1.1 Cafes

Dunbar (2017) in his study claims that food and communal eating has positive benefits for both the individual and society. It fosters more satisfied life and trust and strengthens local communities. Moreover, it is a necessity and if sustainable practices can be infused in food, it can be applied to other aspects of life as well. To create a sustainable community, it is imperative to include food as one of the tools to reach out to more people and garner more attention. The cafe doesn't just offer food, desserts, and beverages, it also practises sustainability in the kitchen and the space through local produce, refurbished furniture and decor, minimal consumption of plastic, refusing the straw at all times, and making people feel positive in general with the food and the ambience. The purpose is to demonstrate that people do not have to compromise on their existing lifestyle but introduce change in a step-by-step way by making little changes which are not only sustainable but also economical in the long run. Various ways in the cafe practises low waste are:

a. Plastic recycles
b. Coffee grounds collected by homegrown brands and customers to make everyday products like soaps, cleaning products, etc.
c. Composting
d. Local produce

4.5.1.2 Stores

The products in the store are carefully and consciously curated and are low waste providing alternatives for a plastic-free lifestyle through daily utility and lifestyle products. The product category includes organic, handmade, recycled, upcycled, and fair trade products.

The brands associated with, practice sustainability on various factors, from procurement to production to packaging. The brands either produce all the products by themselves or have collaborated with another self-help group and local artisan communities and practise fair trade in terms of remuneration. The store has also partnered with various NGOs from time to time for procurement and outsource needs in terms of products, promotional items, and decor items, which were made out of waste materials collected by Greenstraw as a part of the zero waste drive and thus facilitating the circular economy. Initiatives like these go beyond environmental sustainability and include social sustainability as a part of operations by providing better opportunities to such communities, ensuring holistic development of the whole community impacting the overall development of city or country.

4.5.1.3 Coworking and Community Space

The space welcomes people of all interests and fields to come and engage in a dialogue to start a conversation on environmental sustainability. The people who visit are associated with start-ups, freelancers, artists, educators, musicians, and anyone in general who shares the curiosity about sustainable practices and considers embarking on an earth-friendly journey slowly and steadily.

4.5.2 Greenstraw: Ideology and Focus

4.5.2.1 Communication and Community

Past research (Arkowitz & Lilienfeld, 2017; Kahan, 2008) and practical experiences reflect that behaviour change of people is challenging due to lack of awareness, and absence of key information (the consequences, the alternatives, the risks, the benefits, etc.) and apprehensions about making a change towards sustainability. When benefits are discussed, they seem far-fetched, arguments for the behaviour change are ambiguous and theoretical and therefore people do not change in practice. Communication therefore becomes an important link in bringing out a pro-sustainability behaviour change, addressing the apprehensions, and also providing confidence that sustainability does not mean sacrificing the existing lifestyle, but exploring alternate ways to maintain it which does not compromise on the needs of the future generation. Supporting this, Kritika, a partner quipped, "It is communication that changes knowledge, ideation and environment, and once this cycle started people's self-faith grew which reflected in the conviction of their behaviour".

4.5.2.2 Sustainability Focus at Greenstraw

Greenstraw harps upon environmental and social facets of sustainability. For environmental aspects, Greenstraw focused upon reducing food wastage, composting, minimising the use of plastic, and sourcing locally. For e.g. from the inception of the cafe, it has saved approximately 14,000 plastic drinking straws by substituting it with steel drinking straws, and also doing away with PET bottles. It imbibes the essence of a circular economy in their production cycle and logistics and sources locally produced confectioneries, and other items from home chefs and bakers in reusable packaging. Pricing of the menu is at par with the hospitality market, breaking the perception that sustainability is a complex and expensive practice and fulfilling the business's economic aspect.

It provides coworking and community space which showcases events in the themes of sustainability and caters to an audience which is open to know about sustainable practices across genders and age groups. During all the hosted events, there is no plastic waste, and food waste for the excess is distributed amongst the needy. The participants of the events are reminded to be thoughtful before they take their serving by placing small cards which carry thoughtful quotes and sayings appealing to their emotions.

In all the verticals (lifestyle store, cafe, and coworking and community space) Greenstraw attempts to practise inclusivity in terms of the audience and those involved as a part of their operations and commerce along with environmental sustainability. In the store, it is reflected in creators who source/supply the products to the store, where designers and local artisans are featured on one platform, supporting the ethos of coexistence. Moreover, it actively hosts and collaborates for events and workshops that foster inclusivity irrespective of age, gender, and socio-economic background and focus on imbibing a culture of sustainability and ethics in all aspects of life. These workshops prove to be one of the most effective and engaging ways to introduce and incorporate these practices (refer to Appendix 1). The guiding principle is to question and find alternatives to existing life practices that equip the current and future generations to live consciously without compromising on their needs, aspirations, and lifestyle. One of the strategies to imbibe this is by applying Behavioural Drivers Model, which aids to drive change by fostering and encouraging a behavioural change in the stakeholders (Petit, 2019). The case study attempts to apply the BDM conceptual framework to map and direct its operations in the most effective way maintaining the ethos of sustainability and inclusivity.

4.5.2.3 Social Sustainability and Inclusivity in Greenstraw

Social sustainability and inclusivity are practised across all the verticals of Greenstraw (refer to the Appendix 2). It broadly practises social sustainability and inclusivity in the following three ways:

1. **On the basis of gender**

In terms of workshops, curating events, and facilitators, women actively participate more than men (refer to Appendix 2a). In the store, most sustainable brands are majorly owned by women. Joekes's (1999) study reveals that women play an essential role in engaging in activities which are focused on environmental sustainability. In the case of Greenstraw this proves to be true, for it is observed that whenever plantation drives, workshops, etc., have been conducted women have always outnumbered men. Moreover, Family, friends, institutions, and media all exert a deterministic influence on a child's (and future adult's) gender role. The individual background of a person will have a significant influence on how gender is applied to her/his decisions and behaviours; and the overall socialisation process will condition attitudes and multiple social norms. Responsibly transforming the preconceptions rooted in experience and exposure is a key aspect of change (Witt, 1997) and by applying a gender transformative lens, should consider the role of power in social relations, the importance of childhood socialisation, how gender-related norms become embedded in institutions, and how gender norms are produced and reproduced through daily interactions (Marcus et al., 2015).

2. **Education**

Given the community embedding various people are involved as facilitator, collaborator, audience, and even the workforce. Many of these people are first generation to be literate and have received some degree of education. Amongst these, some of them come from families where one parent is literate and in some both the parents are not literate (refer to the Appendix 2b).

3. **Inclusivity as social cause**

a. **Promoting homegrown brands**

The brands at the store associated practice sustainability across levels, from procurement to production to packaging. The brands irrespective of producing the products or sourcing it from collaborated self-help groups, practise fair trade and encourage the marginalised communities' (socio-economically marginalised) participation in the mainstream products.

b. **Collaborations with co-operative society and NGOs**

Greenstraw regularly partners with various NGOs for procurement and outsourcing products, promotional items, decor items, made out of waste materials collected as a part of the zero waste drive and thus facilitating the circular economy. Initiatives like these go beyond environmental sustainability by providing better opportunities to such communities incorporating inclusivity in its operations and ultimately contributing to the development of the country.

4.6 Conclusion

The experiences and spaces that Greenstraw creates act as a facilitator to bring about a sustainable change in people's lifestyle and create positive action and conversation around it. All the initiatives are developed focusing on the intent to bring about systematic social and behavioural change with an aim to influence multiple determinants which can broadly be categorised into the following (Petit, 2019):

- Psychology, gathering individual cognitive and emotional drivers—Greenstraw hosts multiple workshops and events which address the issues and apprehensions of the people about sustainability, and break them by offering solutions which do not compromise on their existing lifestyle.
- Sociology, for determinants related to interactions within families, communities, groups, and society at large;—Greenstraw's events and workshops are curated aiming at larger groups of people irrespective of their age, gender, socio-economic background, and the level of orientation of sustainable practices.
- Environment, for structural elements such as institutions, policies, systems and services, infrastructures, information, etc.—Greenstraw practises environmental sustainability and promotes it by operationalising it across all the verticals.

It is imperative for sustained behavioural change that the environment itself induces positive actions from the target group. Therefore, creating a cohesive environment to bring about the needed change is of utmost importance. Social and behaviour change programming should always consist in multi-faceted strategies and tactics winning people's hearts and minds, but also winning the crowd and shaping the environment to induce positive actions (Bandura, 2001; Bronfenbrenner & Ceci, 1994; Golden et al., 2015; Liu et al., 2017; Pajares, 2002; Rogers et al., 2013).

Greenstraw attempts to infuse a culture of sustainability and inclusivity based on its practice and communication. However, it is at a nascent stage and has just begun to garner some attention by starting a discourse on the same. While there are multiple projects in pipelines like:

1. Development of the school that Greenstraw has adopted in Daskroi district which is a rural part. The project focuses on imparting vocational skill programmes to the students of higher secondary school as well as improving the infrastructure

of the school by adding rainwater harvesting, solar panels, and composting and introducing practical environmental science modules.
2. Provide menstrual hygiene education and awareness in urban slums.
3. Promote and boost the cultural heritage and provide a platform to artisans and craftsmen, by developing art spaces driven by sustainability.

It has also devised a robust future action plan which strengthens the present and future:

1. Reach out to more stakeholders for zero-waste consulting across industries
2. Take the pilot school plan to governments and major CSR funding
3. Expand the concept of the spaces to more cities and add more causes
4. Develop more campaigns for the foundation to connect more social causes to sustainability
5. Creating drivers of change through being one
6. Create a cohesive environment towards sustainability by identifying different groups of people/stakeholders who are drivers in change for a long-term impact.

Being a young enterprise, as well as in the team, Greenstraw is keen to experiment and learn from its experiences and discourses around it by creating a shared experience with the community. The grit and determination of the team towards sustainability and inclusion only increases as each day passes, motivating them to streamline their activities, and map and assess their impact to serve better.

Appendices

Appendix 1

What We Do: Diversity in Inclusivity

1. Events and workshops

We curate and conduct events and workshops to make people aware about sustainable living by providing alternatives to lead the least impact life.

a. Plantation drive/free sapling distribution/plant parenting, etc.
b. Food: Grow your own food/super food/microgreens/drink fermentation/vegan baking workshop/magical millet.
c. Everyday products: Composting/shampoo making/soap making/deodorant making.
d. Building scientific thinking/next gen warriors: renewable energy workshop for kids/awareness campaigns.
e. Cultural events.
f. Heritage walks.

g. Music events.
h. Social cause/drives/meet up.
i. Collaboration—Low waste product pop-up at Shreyas Foundation.
j. Event partner for a preloved sale by Ciceroni.

2. Creating community and network through collaboration/pop-up, etc., in order to bring greater change and make sustainable practices part of collective consciousness.

a. Thrift pop-up at Greenstraw Usmanpura
b. Thrift pop-up at La Maison
c. Thrift pop-up at Hive
d. Product pop-up at Maisha
e. Product pop-up at Ford Plant Sanand
f. Drives: e-waste collection, yoga mats, cigarette butt

3. Other activities

a. Sustainability models

We design sustainability models and guidelines for people, communities, and spaces where they can implement the same to achieve maximum sustainability.

b. Low waste product curation and consulting

We help you switch to a low waste lifestyle through everyday products that are low impact in all ways.

c. Awareness campaigns

We design and implement awareness campaigns by identifying various environmental and social factors that needs to be addressed in our communities.

Appendix 2

2a. Social Inclusion—Mapping of Greenstraw—Gender Basis

For the period 1 April 2022–30 August 2022.

Category	Female	Male	Total
Lifestyle products	45	5	50
Workshop facilitators	6	0	6
Workshop participants (above 18)	46	10	56
Workshop participants (below 18)	5	11	16
Workforce	6	10	16
Event/drive collaborators	7	3	10

2b. Socio-Economic Background

Educated Households

Category	Both parents literate	One literate parent	None	Total
Lifestyle products	50	0	0	50
Workshop facilitators	4	2	0	6
Workshop participants (above 18)	40	16	0	56
Workshop participants (below 18)	9	7	0	16
Workforce	9	0	7	16
Event/drive collaborators	7	3	0	10
School	0	500	200	700

On Basis of Household Income

Category	Daily wagers −300,000	300,000–500,000	Above 500,000	Total
Lifestyle products	0	0	50	50
Workshop facilitators	0	1	5	6
Workshop participants (above 18)	0	0	56	56
Workshop participants (below 18)	0	0	16	16
Workforce	9	0	7	16
Event/drive collaborators	0	0	10	10
School	700	0	0	700

References

Arkowitz, H., & Lilienfeld, S. O. (2017). *Facts and fictions in mental health*. Wiley.

Bandura, A. (2001). Social cognitive theory: An agentic perspective. *Annual Review of Psychology, 52*(1), 1–26.

Bandura, A., & Walters, R. H. (1977). *Social learning theory*. Prentice-Hall.

Bronfenbrenner, U., & Ceci, S. J. (1994). Nature-nuture reconceptualized in developmental perspective: A bioecological model. *Psychological Review, 101*(4), 568.

Brundtland, G. H. (1987, March 20). *Our common future*. United Nations. Available at: https://whc.unesco.org/en/sustainabledevelopment/ (Accessed: October 20, 2022).

Chakraborty, S. K. (1987). *Managerial effectiveness and quality of worklife: Indian insights*. Tata McGraw-Hill. Available at: https://www.cbgaindia.org/research/fiscal-policy-social-inclusion/ (Accessed: October 25, 2022).

Dahlsrud, A. (2008). How corporate social responsibility is defined: An analysis of 37 definitions. *Corporate Social Responsibility and Environmental Management, 15*(1), 1–13.

Das, K. N. (2022). *India's population growth slows as world reaches 8 billion*. Thomson Reuters. Available at: https://www.reuters.com/world/india/indias-population-growth-slows-world-reaches-8-billion-2022-11-15/ (Accessed: December 9, 2022).

Dey, M. (2021). World Environment Day: A look at the history of sustainable architecture in ancient India. *The Financial Express*. Available at: https://www.financialexpress.com/lifestyle/science/world-environment-day-a-look-at-the-history-of-sustainable-architecture-in-ancient-india/2264327/ (Accessed: December 9, 2022).

Dixon, J. A., & Fallon, L. A. (2008). *The concept of sustainability: Origins, extensions, and usefulness for policy* (Vol. 2, March 2015, Hal. 37–41). EMES Network. Social Enterprise. Available at: https://emes.net/focus-areas/social-enterprise/ (Accessed: October 25, 2022).

Dunbar, R. I. M. (2017). Breaking bread. *Adaptive Human Behaviour and Physiology, 3*(3), 198–211.

Giovannoni, E., & Fabietti, G. (2013). What is sustainability? A review of the concept and its applications. In *Integrated reporting* (pp. 21–40). Springer.

Golden, S. D., McLeroy, K. R., Green, L. W., Earp, J. A. L., & Lieberman, L. D. (2015). Upending the social ecological model to guide health promotion efforts toward policy and environmental change. *Health Education & Behavior*, *42*(1_suppl), 8S–14S.

Heyzer, S. J. W. N., & Salles, R. O. O. A. V. (1994). Gender, environment and population. *Development and Change, 25*(1), 137–165.

Jayawardhana, K., Fernando, I., & Siyambalapitiya, J. (2022). Sustainability in social enterprise research: A systematic literature review. *SAGE Open, 12*(3), https://doi.org/10.1177/21582440221123200.

Joekes, S. (1999). A gender-analytical perspective on trade and sustainable development. In *Trade, sustainable development and gender* (pp. 33–59). UNCTAD.

Kahan, D. M. (2008). Cultural cognition as a conception of the cultural theory of risk. In S. Roeser (Ed.), *Handbook of risk theory* (Harvard Law School Program on Risk Regulation Research Paper, 08-20).

Kerlin, J. A. (2010). A comparative analysis of the global emergence of social enterprise. *VOLUNTAS: International Journal of Voluntary and Nonprofit Organizations, 21*(2), 162–179.

Kummitha, R. K. R. (2017). *Social entrepreneurship and social inclusion*. Palgrave.

Liu, X., Zhang, S., & Bae, J. (2017). The impact of renewable energy and agriculture on carbon dioxide emissions: Investigating the environmental Kuznets curve in four selected ASEAN countries. *Journal of Cleaner Production, 164*, 1239–1247.

Manjrekar, S. M. (2017). *Study of sustainability practices ingrained in Indian culture*. 5th International Conference on Civil, Architecture, Environment and Waste Management (CAEWM-17) (Vol. 17758).

Marcus, R., Harper, C., Brodbeck, S., & Page, E. (2015). *Social norms, gender norms and adolescent girls: A brief guide*. Overseas Development Institute.

McClelland, D. C. (1975). *Power: The inner experience*. Irvington.

Mehta, R. (2021). *It's complicated: India's history with sustainability*. The Green Medium. Available at: http://www.thegreenmedium.com/blog/2021/1/3/its-complicated-indias-history-with-sustainability (Accessed: December 9, 2022).

Nations, U. (2016). *The sustainable development goals 2016*. eSocialSciences.

Pajares, F. (2002). Gender and perceived self-efficacy in self-regulated learning. *Theory into Practice, 41*(2), 116–125.

Pandey, A. (2017, October). *What India can teach the world about sustainability* (Vol. 2). World Economic Forum. Available at: https://www.weforum.org/agenda/2017/10/what-india-can-teach-the-world-about-sustainability/ (Accessed: October 25, 2022).

Pandey, N., & Sahay, A. (2022). Social entrepreneurship in India. In *Indigenous Indian management* (pp. 347–383). Palgrave Macmillan.

Petit, V. (2019). *The behavioural drivers model: A conceptual framework for social and behaviour change programming*. UNICEF Middle East and North Africa Regional Office.

Pradeep, S. (2020). *Sustainability lessons from ancient India*. ValuEarth. Available at: https://www.valuearth.com/sustainability-lessons-from-ancient-india/ (Accessed: December 9, 2022).

Rajora, G., & Yallo, K. (2021, August 25). *Understanding social inclusion in India: Perspective from below*. Available at: https://herald.uohyd.ac.in/understanding-social-inclusion-in-india-perspective-from-below/ (Accessed: November 7, 2022).

Rogers, S. H., Gardner, K. H., & Carlson, C. H. (2013). Social capital and walkability as social aspects of sustainability. *Sustainability, 5*(8), 3473–3483.

Sengupta, S., & Sahay, A. (2017). Social entrepreneurship research in Asia-Pacific: Perspectives and opportunities. *Social Enterprise Journal*. https://doi.org/10.1108/SEJ-11-2016-0050

Sharma, R., Aggarwal, N., & Kumar, S. (2014). Ecological sustainability in India through the ages. *International Research Journal of Environment Sciences, 3*(1), 70–73.

Shukla, M. (2011). *'Landscape' of social entrepreneurship in India: An eclectic inquiry* (Working Paper). https://doi.org/10.2139/ssrn.2073234

Shukla, M. (2020). *Social entrepreneurship in India: Quarter idealism and a pound of pragmatism*. Sage.

Social sustainability and inclusion. (2021). The World Bank. Available at: https://www.worldbank.org/en/topic/socialsustainability/overview (Accessed: December 9, 2022).

SUMAS Team. (2021, April 26). *Sustainability 101—Why is social inclusion important?* SUMAS. Available at: https://sumas.ch/sustainability-101-why-is-social-inclusion-important/ (Accessed: November 2, 2022).

United Nations. (2020). *World Youth Report (WYR) for youth*. United Nations. Available at: https://www.un.org/development/desa/youth/world-youth-report.html (Accessed: November 2, 2022).

Vos, R. O. (2007). Defining sustainability: A conceptual orientation. *Journal of Chemical Technology & Biotechnology: International Research in Process, Environmental & Clean Technology, 82*(4), 334–339.

Werbach, A. (2011). Strategy for sustainability. *Strategic Direction*. https://doi.org/10.1108/sd.2011.05627jaa.013

Wilkinson, A., Hill, M., & Gollan, P. (2001). The sustainability debate. *International Journal of Operations & Production Management, 21*, 1492–1502.

Witt, S. D. (1997). Parental influence on children's socialization to gender roles. *Adolescence, 32*(126), 253. Available at: https://www.proquest.com/openview/16ccf7f83986fed6f37ad8be6e8144bb/1?pq-origsite=gscholar&cbl=41539 (Accessed: October 26, 2022).

Yang, W., Roig, M., Jimenez, M., Perry, J., & Shepherd, A. (2016). *Leaving no one behind: The imperative of inclusive development*. Available at: https://www.un.org/esa/socdev/rwss/2016/chapter1.pdf (Accessed: October 25, 2022).

Part II
Managing Sustainability

Chapter 5
Sustainability Reforms and Corporate Disclosure: The Case of a Small Island Developing State

Dineshwar Ramdhony

Abstract The purpose of this study is to investigate how listed companies in Mauritius conceive of the environment and biodiversity and communicate their commitment to the protection of the same in the wake of sustainability reforms that include national projects such as the Maurice Ile Durable (MID) and the imposition of green taxes. Special attention is paid to firms' commitment in their disclosures as to whether they are based on instrumental or non-instrumental values. We rely on a content analysis of disclosures supplemented by discourse analysis of the corporate statements of commitment to interpret the organisational motivations towards the environment and biodiversity. The sample includes all companies listed on the Stock Exchange of Mauritius (SEM) during the nine years from 2008 to 2016. Thematic analysis was carried out using NVivo 12-Plus software. The results show that there is an increase in the level of disclosure relating to the environment. As for the quality of disclosure, the declarative form is preferred by most companies, but over the years, there has been progress towards quantitative disclosures. It was also observed that companies in Mauritius had kept a strong anthropocentric paradigm in their environmental disclosures over the nine years. While the launch of the MID project did not result in any observable change in the extent of disclosures, the doubling of MID levy on liquefied petroleum gas is associated with an increase in environmental disclosures. Thematic analysis showed that themes such as "awareness campaigns", "carbon footprint", "clean-up campaigns", "eco-friendly alternatives" and "environmental policies/regulations" started occurring more often in annual reports from 2012 while disclosures about "climate change" experienced a rise only from 2016. This study provides new insights on environment and biodiversity reporting in a developing economy context and Small Island Developing States (SIDS).

Keywords Biodiversity · Content analysis · Quality of disclosure · Thematic analysis · Instrumental values · Non-instrumental values

---

D. Ramdhony (✉)
University of Mauritius, Moka, Mauritius

D. Crowther and S. Seifi (eds.), *Preparing for a Sustainable Future*,
Approaches to Global Sustainability, Markets, and Governance,
https://doi.org/10.1007/978-981-99-2456-1_5

5.1 Introduction

Reporting practice has significantly evolved since the 1970s whereby social reports have complemented traditional financial reporting (Cormier & Gordon, 2001) and, since the 1980s, by environmental reports (Cormier et al., 2005). Such environmental disclosures are often described as accounting narratives due to their more verbal nature as opposed to numerical information (Cho et al., 2010). This evolution towards environmental accounting narratives may be attributed to companies now being expected to operate by considering the economic, social and environmental issues prevailing at the wider societal level. It is expected that companies conduct their business responsibly and provide environmental disclosures that are actively scrutinised (Radu & Francoeur, 2017). Environmental accounting practice has received attention from researchers worldwide due to the increase in public concern for environmental issues (Eltaib, 2012).

However, it must be acknowledged that there is systematic variation in the quantity and quality of environmental disclosure depending on the nature of a firm's business activities, its environmental performance, organisational size, financial resources, media visibility along with both ownership and board composition (Baalouch et al., 2019; Cormier & Magnan, 2003; Cullen & Christopher, 2002; Patten, 2002b). Quality of disclosure is understood as the precision, relevance and usefulness of the reported information (Cormier et al., 2005).

Theoretically, it is conceived that the organisational motivation for disclosing environmental information is stakeholder-theory led. At one end, an emphasis on the human condition is the key consideration to satisfy and protect; in other words, protecting the environment and biodiversity is valuable to meet human needs. On the other end, the corporate action's emphasis is on the environment itself as the key stakeholder. As such, this study's first objective is to assess whether listed companies in Mauritius pay importance to the environment and biodiversity and communicate their commitment to the protection of the same based on instrumental or non-instrumental values. The instrumental value of biodiversity is dependent on the extent to which it is useful for humans (Gaia & John Jones, 2017; Justus et al., 2009). On the other hand, non-instrumental values are based on the argument that the non-human world and biodiversity are valuable for their own sake (Gaia & John Jones, 2017).

While there is prior evidence to suggest that companies in Mauritius have been increasing their level of disclosure concerning corporate social responsibility and on the environment and that the volume of disclosure does vary in relation to typical organisational variables (Ramdhony, 2017), our contribution lies in revealing the qualitative characteristics of these disclosures and the extent to which companies are embedding instrumental motivations or non-instrumental ones. We are motivated to study the context of Mauritius due to its image positioning as a touristic destination with a major concern about maintaining its natural features and a political agenda to adopt sustainable development policies at the national level. In addition, corporate environmental reporting studies are often carried out in countries where these disclosures are mandatory (Alipour et al., 2019), thus making Mauritius a unique and interesting research context.

Hence our main research question is "How do listed companies in Mauritius conceive of the environment and biodiversity and communicate their commitment to the protection of same?".

One important factor that may cause a change in environmental disclosure in Mauritius regarding biodiversity values is a national project, Maurice Ile Durable (MID). The MID concept, adopted in 2008, is a development strategy towards stimulating sustainable development. It is described as the "new sustainable vision to guide national development" (National Report of the Republic of Mauritius, 2015). Five broad sectors, the 5Es, have been identified as the areas of focus: Energy; Environment; Employment and Economy; Education; and Equity. Sustainable development is referred to as "encouraging economic growth while protecting our environment/ecology and improving our quality of life – all without affecting the ability of future generations to do the same" (Presentation on Maurice Ile Durable United Nations in New York 1 November 2013). Big groups of companies such as Terra, Omicane and Alteo are operating in the cane industry and have a deep impact on diverse sectors such as agriculture, energy, environment, waste management and carbon emissions. Thus, they must disclose environmental information that caters simultaneously for both human and non-human interests. Against this backdrop, the study attempts to answer a second research question? "What environmental/sustainability themes are listed companies emphasising in their annual reports?".

The next section explains the research context. Section 5.3 provides the theoretical framework for this study. A review of the literature is shown in Sect. 5.4. The methodology used for this research is explained in Sect. 5.5, while Sect. 5.6 displays the results. Discussion of the results is found in Sect. 5.7. Finally, Sect. 5.8 provides conclusions to this study.

5.2 Research Context—Mauritius and the Environment

Mauritius is an island that is densely populated, isolated geographically and not in receipt of significant natural resources such as oil, gas and other precious minerals. However, the country is one of the ten most beautiful islands in the world (Eade, 2014). It is no surprise that tourism has emerged as an economic pillar. In 2018, 1,399,408 tourists visited the island, and this number is expected to increase by 3.6% in 2019 (Statistics Mauritius, 2018). With a relatively small size of only 144 square miles, there is pressure on land demand to build more hotels and engage in activities to diversify the economy. The downside of economic development in terms of air, water and noise pollution is a challenge for stakeholders to manage (Mahadeo et al., 2011). To this end, the Report on Corporate Governance for Mauritius (National Committee on Corporate Governance, 2004) states that:

> Environmental issues are therefore particularly important in Mauritius and companies need not only be aware of the importance of these issues but also be actively involved in managing their activities so as to minimise any negative impact on the environment.

Climate change is having a profound impact on the fauna and flora of Mauritius. The Mauritius Meteorological Services has observed lengthening of dry days, heavier rains and stronger cyclones (Mahadew & Appadoo, 2018). Extreme temperatures are being recorded during summer, and flash floods disrupting civilians' lives, creating irrational fears as summer approaches. Sandy beaches used as the main marketing icon to attract travellers to the island are gradually eroded. While the effects of climate change have an impact on tourism arrivals, the reverse has also been proved (Mahadew & Appadoo, 2018).

Mauritius is also prone to natural disasters and facing issues similar to other SIDS, such as total or near depletion of their natural resources (Briguglio, 1995) and rising sea levels endangering their existence. The environment is of critical importance for Mauritius, and businesses thus have the responsibility to care for the environment in which they are operating.

5.3 Theoretical Framework

While various theories have been used to explain the growing attention to environmental disclosures, the stakeholder theory remains one of the most prominent theories (Giacomini et al., 2020). The way in which the relationship between firms and their stakeholders is organised forms the basis of the stakeholder theory (Gray et al., 1995). Freeman (1984) defines stakeholders as "groups who can affect or are affected by the achievement of an organisation's purpose". According to Clarkson (1995), Stakeholders can be categorised as primary and secondary stakeholders. The survivability of a corporation is dependent on the continued participation of primary stakeholders, while secondary stakeholders are not essential to the corporation's survival.

A firm may have several stakeholders. The objectives of these stakeholders may diverge. For instance, one stakeholder may privilege the protection of the environment while another may have respect for human rights at heart. The managerial branch of the stakeholder theory argues that the firm will attend to the demand of powerful stakeholders (Ullmann, 1985). Amid the diverging interests of stakeholders and the power they command, a firm's success depends on the successful management of its relationship with stakeholders (Rashid, 2015). Some stakeholders are so powerful that complying with their requirements becomes essential for the firm's success (Islam & Deegan, 2008) and may even lead to corporate failure if their demands are not met (Deegan, 2002).

Environmental reporting becomes a dialogue between the firm and its stakeholders by providing information about the impact of business activities on the environment to respond to stakeholders' expectations (Barako & Brown, 2008; Gray et al.,

1995; Van der Laan Smith et al., 2005). Stakeholder theory has been widely used in the accounting literature. For instance, Gia and Jones (2017) explain that UK local councils disclose biodiversity information to those stakeholders who can play an important part in biodiversity conservation. Kiliç (2016) used stakeholder theory and legitimacy theory to explain banks' environmental and social reporting in Turkey. Several corporate disclosure studies use the stakeholder theory to explain how corporate disclosure influences major stakeholders' perceptions (Islam & Deegan, 2008; Kamal & Deegan, 2011; Rashid, 2015).

Businesses use the environment as a free resource to carry their activities. In this process, the environment is inevitably harmed. Therefore companies have the moral duty to inform society about the negative impact of their actions on the environment and the initiatives taken to curb the effect of these activities. Stakeholder theory argues that companies can use environmental reporting to discharge their accountability towards stakeholders (Deegan, 2002).

In line with the stakeholder theory, the neo-institutional theory states that firms may mimic what other firms in the industry are reporting. DiMaggio and Powell (1983, p. 152) argue that "organizations tend to model themselves after similar organizations in their field that they perceive to be more legitimate or successful". According to Rahaman et al. (2004), uncertainty may be a motivator for such imitation. As a result, corporations aim to replicate other successful organisations in order to lessen uncertainty.

5.4 Literature Review

5.4.1 Motivations for Environmental Reporting

Disclosures are usually used to reduce the information asymmetry between managers and investors (Cormier et al., 2005). Some investors are looking to invest in environment-conscious companies; therefore, managers will ensure that their companies look attractive to those investors by indulging in environmental reporting. Also, firms with superior environmental performance in industries that negatively impact the environment are incentivised to disclose environmental information to keep stakeholders informed of their strategies (Cormier & Gordon, 2001). The disclosures can provide a more positive view of the organisation to its stakeholders, possibly resulting in better business relationships and better financial performance. A significant association of corporate environmental disclosure with financial performance is revealed by the majority of empirical research (Botosan & Plumlee, 2002).

Disclosure practices can also be viewed from a legitimacy point of view as environmental disclosures can be employed to improve firms' reputation vis-a-vis stakeholders and achieve legitimacy (Hassan & Romilly, 2018). One possible reason for a lower level of environmental reporting could be the lack of accountants' exposure towards environmental reporting (Nazli Nik Ahmad & Sulaiman, 2004). Such a line of thought makes sense as accountants and financial professionals are mostly qualified in financial matters rather than non-financial matters, such as the environment. The cost of gathering information to provide disclosures also discourage companies from indulging in disclosure practices. Besides, disclosing information results in unpredictability and loss of control depending on how external parties react to such information (Gray et al., 1993).

5.4.2 *Biodiversity Values*

The Earth's plethora of species, essential to its well-being, is referred to as biodiversity (Houdet & Germaneau, 2014). Humans can value biodiversity for its contribution to supporting life on Earth, the satisfaction they get from it and its economic benefits, among others (Gaia & John Jones, 2017). These values can be distinguished into strong anthropocentric (shallow), weak anthropocentric (intermediate) and ecocentric (deep) paradigms (Sylvan & Bennett, 1994). Shallow paradigms reject biodiversity's intrinsic value, even though they recognise that biodiversity has instrumental value to human beings. Intermediate paradigms attribute greater value to humans but still recognise that some elements of nature have intrinsic value.

In contrast, deep paradigms assign intrinsic value to biodiversity. This study uses environmental paradigms identified by Connelly and Smith (2003). These values are also used by Gaia and Jones (2017) to investigate whether there is an alignment between these philosophies and the biodiversity conservation explanations of UK local councils. Table 5.1 illustrates the 14 categories of biodiversity.

5.4.3 *Quality of Disclosure*

Quality of disclosure refers to the precision, relevance and usefulness of the reported information (Cormier et al., 2005). As such, results relating to solely the quantity of disclosures omit the aspects of precision, relevance and usefulness. Thus, classifying disclosures as declarative, quantitative and economic allows us to investigate the quality of those disclosures and their suitability to corporate information users.

Most studies that attempt to classify environmental disclosures in such a way find that companies disclose about the environment primarily in a declarative fashion (e.g., Ahmad & Haraf, 2013; Kuasirikun & Sherer, 2004; Nazli Nik Ahmad & Sulaiman, 2004; Thompson & Zakaria, 2004). Nazli Nik Ahmad and Sulaiman (2004) report that 97% of disclosed environmental information is declarative, 2% in non-monetary terms and 1% in monetary terms. Kuasirikun and Sherer (2004) along with Thompson and Zakaria (2004), observe a similar trend with the highest proportion of disclosures being in narrative form, distantly followed by non-monetary and monetary. Ane (2012) also finds that most companies disclose environmental information only

Table 5.1 Biodiversity values

Strong anthropocentric (shallow) paradigms	
Economic value	Biodiversity provides a variety of materials to industries such as agriculture, tourism and construction (Gaia & John Jones, 2017)
Subsistence value	Goods (e.g., food) and services (e.g., natural energy) that are required to fulfil basic human needs are provided by biodiversity (Gaia & John Jones, 2017)
Life supporting value	Biodiversity provides clean air, soil and water, thereby supporting the functions provided by ecosystems services, upon which human beings' survival is dependent (Gaia & John Jones, 2017)
Recreational value	Outdoor recreational opportunities (e.g., fishing, hiking, wildlife photography) are provided to humans by biodiversity (Gaia & John Jones, 2017)
Therapeutic value	Given that the natural environment is ideal for treating both mental and physical ailments, biodiversity contributes to the well-being of humans (Gaia & John Jones, 2017)
Educational value	Learning processes (e.g., risk calculation) are stimulated, and virtues (e.g., serenity) are developed by nature (Gaia & John Jones, 2017)
Aesthetic value	Contact with parts of nature that humans consider aesthetically pleasing and delightful can provide them with enjoyment (Chapin et al., 2000; Gaia & John Jones, 2017)
Scientific value	Scientific discoveries stem from the library of biology, curiosity and inspiration that biodiversity offers (Gaia & John Jones, 2017)
Spiritual value	Due to the religious attribute of nature (e.g., cattle in Hinduism), humans are willing to preserve biodiversity (Gaia & John Jones, 2017)
Cultural symbolisation value	The parts of nature that provide cultural significance and remembrance of personal experiences induce a feeling of moral obligation in humans (Gaia and Jones, 2017)
Historical value	Natural places where historical events (e.g., battle sites) took place might hold historical values (Gaia & John Jones, 2017)
Weak anthropocentric (intermediate) paradigms	
Future value	The need to preserve biodiversity such that future generations have a natural environment as varied and rich as the current one is referred to as future value (Gaia & John Jones, 2017; Jones, 2003)

(continued)

Table 5.1 (continued)

Instrumental value to non-human things	Living things other than humans (e.g., animals) are supported by biodiversity, hence making it valuable (Gaia & John Jones, 2017)
Ecocentric (deep) paradigms	
Intrinsic value	Irrespective of its usefulness to humans, biodiversity holds value by virtue of its existence (Gaia & John Jones, 2017)

in qualitative terms, with 78% of disclosures being in qualitative terms, while the remaining 22% of information is disclosed in quantitative terms.

As opposed to previous findings, Fatima et al. (2015) report that the environmental information disclosed by the companies on the stock exchange of Malaysia is mostly quantitative. However, such information is still reported primarily in non-monetary terms. The results of Kabir and Akinnusi (2012) are also not in line with the literature. Even though the study finds that most environmental disclosures were in descriptive form, there was a more considerable amount of information disclosed in monetary terms than non-monetary terms. Nazli Nik Ahmad and Sulaiman (2004) infer that the environmental disclosures are more of a public relations tool used to enhance the image and reputation of companies due to the lack of information in monetary terms that would be most useful to users of corporate information.

5.5 Research Methodology

5.5.1 Data Collection and Sample

We rely on a content analysis of disclosures supplemented by discourse analysis of the corporate statements of commitment to interpret the organisational motivations towards the environment and biodiversity. Scholars who study social and environmental reporting have extensively employed content analysis (see Milne & Adler, 1999; Patten, 2002a, 2002b; Chu et al., 2013), which has proven to be a useful technique for the collection and evaluation of data from voluntary disclosures (Adler et al., 2017). Fairclough (1992) argues that any discourse (written or spoken) can be seen as being simultaneously a piece of text, an example of discursive practice and one of the social practices.

The sample used includes all the companies listed on the Stock Exchange of Mauritius during the nine years from 2008 to 2016. We choose 2008 as it coincides with the beginning of reforms to make Mauritius a "green island" (See MID p3). As companies may take time to adjust their disclosures to reflect new and amended environmental practices, we analyse the disclosures over nine years. Annual reports were the only data source used for this project.

5.5.2 Content Analysis

To determine the quality of disclosure, the study employs categorisation based on whether environmental and biodiversity information is disclosed in declarative terms, quantitative terms or monetary terms. Such a methodology has been widely used in prior studies (e.g., Ahmad & Haraf, 2013; Kuasirikun & Sherer, 2004; Nazli Nik Ahmad & Sulaiman, 2004; Thompson & Zakaria, 2004). For example, if a company discloses environmental and biodiversity information in declarative terms in a particular year, it obtains a score of 1 in that category. If the company does not disclose environmental and biodiversity information in that category, a score of 0 is given instead. The sum of all the scores obtained by companies each year is then calculated for each category to obtain the total content analysis score of that category.

Environmental disclosures are classified on a spectrum of 14 biodiversity values based on the methodology used by Gaia and Jones (2017). The score given to each value is based on the number of sentences that relate to that value. According to Milne and Adler (1999), using sentences for both coding and measurement provides "complete, reliable and meaningful data for further analysis". The shallow paradigm includes Economic value, Subsistence value, life-supporting value, Recreational value, Therapeutic value, Aesthetic value, Scientific value, Spiritual value, Cultural symbolisation value, Historical value and Educational value. The intermediate paradigms include Future value and Instrumental value to non-human things, while deep paradigms consist of all narratives relating to biodiversity's intrinsic value. This allows for discourse analysis to be carried out to investigate whether listed companies view the environment based on instrumental or non-instrumental values.

5.5.3 Thematic Analysis

Data was gathered from the disclosure statements of 37 companies for the years 2008 to 2016, extracted from their annual reports. A first overview of the data was given by reviewing the disclosure statements and recording the number of pages for each statement obtained for the specified years in an Excel sheet.

As the data was already in electronic format, it was entered into a qualitative data analysis software called NVivo 12-Plus for thematic analysis. The data was classified year-wise, implying that folder 1 in NVivo contained the disclosure statements for the Year 2008, folder 2 would refer to disclosure statements for the Year 2009 and folder 3 for the Year 2010. Hence, there was a total of 9 folders, separating the disclosure statements by year. Themes related to the environment and sustainability were then coded through an initial broad coding strategy with an initial node. Using "Coding Queries" in NVivo, the main themes appearing were recorded: Awareness campaigns, Carbon footprint, Clean-up campaigns, Climate change, Eco-friendly alternatives, Endangered Species Conservation, Environmental Policies/Regulations,

Protection for the environment, Greenhouse Gas (GHG) Emission, Global warming, Green practices, "Maurice Ile Durable" (MID) Project, Minimising environmental impacts, Paper recycling, Tree planting, Reducing paper consumption, Renewable energy, Solar panels, Sustainable development and Waste Recycling (batteries, tyres, oil, …).

5.6 Results

Table 5.2 shows that environmental disclosures by companies listed on the SEM are mostly in declarative terms followed distantly by quantitative and monetary terms. These results align with previous studies (e.g., Ahmad & Haraf, 2013; Kuasirikun & Sherer, 2004; Nazli Nik Ahmad & Sulaiman, 2004; Thompson & Zakaria, 2004) in terms of the proportion of environmental disclosures in each category. It emphasises that environmental information is mainly disclosed in very general terms, without specifying monetary amounts, in annual reports (Nazli Nik Ahmad & Sulaiman, 2004). However, an increasing trend in quantitative information is noted post-2011, linked to the CSR levy. Firms were required to spend 2% of their profit on CSR activities. As such, it appears that they have disclosed those activities.

Table 5.3 makes it clear that listed companies in Mauritius lay importance on environment and biodiversity and communicate their commitment to the protection of same based mostly on instrumental values as the percentage of sentences relating to shallow paradigms is highest in 2008 at 82.9%, 2012 at 81.2% and 2016 at 82.2%. Subsistence value, which relates to the provision of goods (e.g., food and raw materials) and services (e.g., natural energy) for the fulfilment of basic human needs (Gaia & John Jones, 2017), was the value most disclosed for about every year during the period studied. There is insignificant variation in the proportion of each paradigm over the years in terms of narratives relating to shallow, intermediate and deep paradigms. Additionally, over the nine years under study, not a single disclosure about the intrinsic value of biodiversity was made.

Table 5.2 Quality of disclosure

Year	Declarative (%)	Quantitative (%)	Monetary (%)
2008	83.33	8.33	8.33
2009	74.36	15.38	10.26
2010	92.31	2.56	5.13
2011	83.33	0.00	16.67
2012	77.63	10.53	11.84
2013	82.29	8.33	9.38
2014	71.70	24.53	3.77
2015	63.33	27.33	9.33
2016	61.40	32.46	6.14

Table 5.3 Evolution of biodiversity narratives

Biodiversity values	Number of sentences			% of total sentences		
	2008	2012	2016	2008	2012	2016
Economic value	0	4	11	0.0	2.9	8.5
Subsistence value	31	99	79	75.6	71.7	61.2
Life supporting value	0	1	0	0.0	0.7	0.0
Recreational value	0	3	6	0.0	2.2	4.7
Therapeutic value	0	0	0	0.0	0.0	0.0
Aesthetic value	0	0	1	0.0	0.0	0.8
Scientific value	0	0	0	0.0	0.0	0.0
Spiritual value	0	0	0	0.0	0.0	0.0
Cultural symbolisation value	0	0	0	0.0	0.0	0.0
Historical value	0	2	3	0.0	1.4	2.3
Educational value	3	3	6	7.3	2.2	4.7
Total shallow paradigms	*34*	*112*	*106*	*82.9*	*81.2*	*82.2*
Future value	0	3	4	0.0	2.2	3.1
Instrumental value to non-human things	7	23	19	17.1	16.7	14.7
Total intermediate paradigms	*7*	*26*	*23*	*17.1*	*18.8*	*17.8*
Intrinsic value (Deep paradigms)	0	0	0	0.0	0.0	0.0
Total narratives on biodiversity's importance	*41*	*138*	*129*	*100.0*	*100.0*	*100.0*

According to Table 5.4, listed companies in Mauritius more than halved their environmental disclosures after the introduction of the CSR levy in 2010. A possible reason is the community focus of the CSR levy. However, the introduction of a double MID levy and other environmentally motivated taxes in 2011 resulted in a surge in environmental disclosures from a total number of 12 biodiversity-related sentences in 2011 to 138 in 2012. The number of disclosures thereon kept increasing until 2015. There was a decrease in the number of biodiversity-related sentences from 246 in 2015 to 129 in 2016.

Table 5.5 shows themes favoured by companies listed on the stock exchange of Mauritius during the periods 2008–2012 and 2012–2016. It can be seen that over the years, there is an increase in disclosure for all of the seven favourite themes.

5.7 Analysis and Discussion

The content analysis results show an increase in the level of disclosure concerning environmental information. This is consistent with the findings of Gerged (2021), who also finds an increasing trend in corporate environmental disclosures for companies in an emerging market (i.e. Jordan). A remarkable example of this is Terra

Table 5.4 Initiatives and events

Year	Number of biodiversity-related sentences	Initiative and event
2008	41	Introduction of Maurice Ile Durable
2009	40	Government imposed three taxes on vehicle ownership, all related to engine size, as measured by cylinder capacity in cubic centimetres (cc). These include a one-off excise duty on the car price of 55% if the engine capacity is less than 1600 cc or 100% if the engine capacity is greater than 1600 cc
2010	39	Introduction of CSR levy
2011	12	Doubling of MID levy on liquefied petroleum gas. Other environmentally motivated taxes providing a more minor contribution to revenue include excises on plastic products (bottles, bags and cans), an "environmental protection fee" (a tax on hotels, guest houses and tourist residences), and a "solidarity levy", which each contributed 0.2 to 0.3% to total tax revenue, and a "passenger fee" which contributed 1.2% to total tax revenues.
2012	138	
2013	135	
2014	170	
2015	246	
2016	129	

Table 5.5 Evolution of themes favoured over the years

Theme	Evolution (2008–2012–2016)
Awareness campaigns	Little mention (2008) to medium mention (2012 onwards)
Carbon footprint	These terms mostly appear from 2012. No mention was made during the year 2008
Clean-up campaigns	
Climate change	Climate change issues are more prevalent during 2016 compared to 2008 and 2012
Eco-friendly alternatives	Very little mention was made in 2008, but the term became more and more common from 2012
Endangered Species Conservation	This theme is increasingly mentioned throughout the years
Environmental Policies/Regulations	More companies include environmental policies from 2012

Mauricia Ltd disclosing 49 environmental narratives in 2015 against 4 in 2008. The extent of disclosures has also been varying differently based on industry. This may be due to different industries having a different inherent environmental impacts, visibility of environmental issues and degree and type of regulatory intervention (Heras-Saizarbitoria et al., 2015). Admittedly, Omnicane Ltd, which operates in the cane industry in Mauritius, impacts diverse sectors such as agriculture, energy, wastewater management and carbon emissions and has a large number of disclosures, for instance, 40 in 2015. On the other hand, newer manufacturing industries and the service sector have significantly lower environmental impacts and fewer visible ecological issues. Therefore, this can explain the lesser degree of disclosure activism in these industries, such as MCB Group Limited (Bank) having only three environmental disclosures in 2015.

As for the quality of disclosure, the declarative form is mostly preferred by companies, but there is a shifting trend towards quantitative disclosures over the years, especially in most recent years. For instance, in 2014, MCB Group Limited disclosed "*the Bank launched the 2nd edition of the preferential credit facilities, named 'Green Loans', in support of projects aiming to save energy and reduce carbon emissions. An investment grant of 8% of the loan amount is offered to the client for investments in 'green' projects that can be 100% financed by the Bank*". Simultaneously, in 2016, the chemical reduction, liquefied petroleum gas saved, electricity saved, diesel saved, heavy fuel oil saved, clean water saved, wastewater reduced and CO_2 emission reductions from the green loans were all disclosed in quantitative form. However, during all of the nine years, disclosures about the environment and biodiversity in monetary terms such as "*Rs 2.24 M redeemed on electricity savings thanks to LED bulbs and Rs 4.4 M saved on gas due to Solar Systems*" (Sun Resorts Limited, 2015) have remained relatively low. The companies listed on the SEM should provide more quantitative and especially monetary term information to stakeholders as these types of information are deemed more useful and of higher quality.

Environmental disclosures have been classified on a spectrum of 14 biodiversity values. It can be seen that companies in Mauritius have kept a strong anthropocentric paradigm in their environmental disclosures over the nine years. In other words, more than 80% of the biodiversity values of environmental disclosures of Mauritian companies in 2008, 2012 and 2016 are from the category of shallow paradigms. Disclosures from this paradigm recognise that biodiversity has instrumental value to human beings but no intrinsic value. For instance, "*MCB Ltd has adopted the Equator Principles, which stand as the governing foundation of the Bank's Environmental and Social Policy in support of its 'responsible financing' of projects*" (MCB Group Limited, 2014) portrays the company as being more responsive to stakeholders rather than the protection of the environment for its intrinsic value. Companies from sectors that affect the environment more such as Air Mauritius, which belongs to the travel sector, also conceive of the environment and its protection similarly. "*The Group also takes account of its responsibilities to other stakeholders including its employees, its customers and the communities affected by its operations, as well as having regard to the impact its business has on the environment*" (Air Mauritius Ltd, 2016). Similarly, it can be seen that Air Mauritius Limited also attributes value to the environment and biodiversity based on its usefulness in making it more accountable to its stakeholders rather than the intrinsic value of biodiversity itself.

Though significantly lower than shallow paradigm disclosures, there have been a stagnating number of intermediate paradigm disclosures over the years, especially from more prominent companies. As an illustration, by reporting "*When we look to the future, we see a world in which Development is Sustainable. Where every individual has the opportunity to succeed, without compromising the needs of the next generation*" in 2014, Terra Mauricia Ltd emphasises the future value that it attributes to the environment and biodiversity. Over nine years, not a single narrative focuses on the intrinsic value of biodiversity, that is, narratives belonging to the deep paradigm. It shows that the conception of companies about the environment and biodiversity has not evolved. In order words, human interests still prevail in case of conflict with non-human interests.

Initiatives and events that might have affected the level of environmental disclosure by listed companies in Mauritius are also investigated. It can be seen that, while the introduction of MID project did not result in any significant change in the extent of disclosures, the doubling of MID levy on liquefied petroleum gas has resulted in an increase in the extent of environmental reporting by 126 sentences (Table 5.4). This noteworthy increase can also be attributed to the other green taxes introduced in that year. These include the excises on plastic products, environmental protection fee, solidary levy and passenger fee. From that point in time, the number of sentences relating to environmental disclosure remained above the 125 thresholds, even reaching an astonishing 246 sentences in 2015. Thus, it can be inferred that initiatives and events, especially those by the government, can help promote disclosure practices regarding the environment. According to Sun et al. (2019), powerful social actors such as the government can shape corporate environmental behaviour. Another observation is the decrease in disclosures following the CSR levy. It seems that companies have had difficulties in adhering to the newly introduced levy and choose to reduce their level of disclosure about the environment due to the focus of the levy being community-based.

While the favourite themes of disclosure of listed companies in Mauritius have all experienced a rise in occurrence over the years, themes such as "awareness campaigns", "carbon footprint", "clean-up campaigns", "eco-friendly alternatives" and "environmental policies/regulations" started occurring increasingly from 2012. On the other hand, disclosure about "climate change" experienced a rise later only in 2016. The increase in environmental disclosures by companies is commendable as it is a sign that companies are increasingly recognising the importance of the environment and biodiversity along with the protection of the same.

The study finds an increase in reported "awareness campaigns" from 2012. Balanced decision-making that is made on the basis of economic, social and environmental welfare for both current and future generations is a requirement for ecologically sustainable development (Medallon & Gallardo, 2014). The following quotes from annual reports evidence this.

"Regular awareness and education exercises are carried out in the various hotels, involving team members and guests, and the company is also involved with the

Global Reporting Initiative at the corporate level, to reinforce its sustainable development commitment" (Lux Island Resorts Ltd, 2013). "Reinforced human capital, based in part on improved employee awareness of environmental issues and the professionalisation of the different trades that IBL brings together" IBL Ltd (2016). "Sensitisation programmes are now ongoing with a view to educate inhabitants on environmental issues and on the importance of maintaining a cleaner and greener living environment" (ENL Land Ltd, 2010).

Through such reported awareness campaigns at the corporate and national levels, firms listed on the SEM can signal to society that they value the environment. In turn, society allows them the right to operate.

Though inexistent during the 2008–2012 period, carbon footprint disclosures become a recurring theme from 2012. Lux Island Resorts Ltd reported in 2013 that their latest project Tread Lightly had been purposely set up to ensure a 100% offsetting of carbon emission during guests' stay. While a company that produces no direct carbon emission might report this fact as an achievement, through its indirect emission, e.g., buying electricity from a coal-fired power station, the company's activities may still be very harmful to the environment. It can even be said that its contribution to climate change is substantially falsified (Schmidt & Meyer, 2009). Therefore, companies should better report on both their direct and indirect emissions to not hinder their legitimisation process when it comes to carbon footprint reporting.

Another theme that followed a similar trend to carbon footprint is clean-up campaigns. Maintaining the environment is essential for our future generations to benefit. To show the public that they care for the environment, companies have disclosed their actions towards a cleaner living environment. Some examples are:

"Several clean-up exercises of public areas were carried out throughout the year by the Rogers group employees and the latest clean-up exercise was done on the beach from Rivière des Gallets to Bel Ombre" (Rogers & Co Ltd, 2015). "ENL Foundation had initiated a clean-up campaign in Cité St Louis/La Butte and Alma region and this initiative had received the full support of the inhabitants of those regions" (ENL Land Ltd, 2012).

Companies listed on the SEM can be considered late when reporting on climate change as disclosures related to that theme were more prevalent only during 2016. Mauritius Eagle Insurance Co. Ltd reported in 2016 that it aims to fight climate change by "focusing on areas such as controlling air conditioning; switching off of lights in areas where not required after office hours; installing passive infra-red lighting in certain common areas; monitoring the purchase of low energy consumption equipment; intensive use of emails to reduce the use of papers; turning off the computers when going out for lunch; and paperless office and rector/verso document printing if required". Corporate disclosures on climate change may serve symbolic rather than substantive purposes (Milne & Grubnic, 2011). If improvement of their reputation and legitimisation of their activities are included in their goals, companies are more likely to adopt climate change disclosures. In fact, some companies acknowledge that their very existence is at stake with climate change. For example, Phoenix Beverages Ltd reported in 2016 that "Climate change remains a threat to crop yields for raw materials such as barley, malt and hops in producing countries".

A term that was rarely mentioned in 2008 but became common from 2012 is eco-friendly alternatives. "The company wishes to go further in the strengthening and affirmation of Le Caudan Waterfront's identity as an eco-friendly destination by building on several ad hoc "green" initiatives that have been taken over a certain period of time, like the use of eco-friendly and biodegradable detergents when it comes to the cleaning of the premises and recycling of used oils among others" (Caudan Development Ltd, 2015). Companies may view such green development as not only regulations that need to be complied with but rather as market opportunities with the rise of environmentalism among consumers (Cheah & Phau, 2011). The rise in eco-friendly alternative disclosures may indeed be an attempt by listed companies to increase their range of potential customers, thereby increasing financial benefits.

Endangered species conservation has been disclosed, steadily over the whole period studied. According to the annual report of Air Mauritius Ltd in 2016, the company has signed the Buckingham Palace Declaration to fight against the traffickers of endangered animals. Additionally, ENL Commercial Ltd states in the same year that they support several NGO initiatives in favour of the protection of endangered endemic species such as sea turtles and birds. According to Stokes et al. (2010), reporting on endangered species is often seen as a flagship of biodiversity reporting. While compared to the previous theme that could directly increase financial benefits, endangered species reporting is better explained by the concept of legitimacy. By showing that endangered species are included in their business strategies and reporting, the companies will be perceived as more environmentally friendly by society, increasing their legitimacy.

The seventh most favoured theme by companies listed on the SEM is environment policies/regulations, which became more prevalent from 2012. In 2015, CIEL Limited disclosed in their annual report that the company established the Environmental and Social Committee to assist its board of directors with meeting its responsibilities with regard to the company's sustainability policies and practices. Innodis Group reported in 2016 that they have developed and implemented social, safety, health and environmental policies and practices to meet existing legislative and regulatory frameworks. All production units are operated in such a manner as to minimise damage to the environment and the neighbourhoods. As stated by Courtland (2010), companies disclose their environmental policies in response to demands by environmentally conscious stakeholders. In this case, disclosure constitutes a strategy to meet the demands of stakeholders who are increasingly interested in knowing what policies and regulations are being put in place regarding the companies' impact on the environment.

5.8 Conclusions

The study investigates the values on which Mauritian companies base themselves when disclosing environmental information and the extent to which disclosures are declarative, quantitative and in monetary terms. The themes of disclosure favoured by Mauritian companies and the extent of changes in sustainability practices and disclosure of Mauritian companies brought by events and initiatives are also explored.

The discourse analysis carried out in the previous section reports that the quality of disclosures related to the environment and biodiversity has improved in recent years with a lower percentage of declarative form disclosures and a higher percentage of quantitative form. It can thus be inferred that companies are willing to provide information that is deemed more useful to stakeholders. As for the values on which companies base themselves when disclosing about biodiversity, it can be seen that the trend has been a stagnating one, with most disclosures belonging to the shallow paradigm. As such, companies listed on the SEM have long kept the habit of considering biodiversity based on its usefulness to humans rather than attributing intrinsic value to biodiversity itself.

Our findings also allow the inference that events and initiatives such as MID may help promote transparency of information about the environment. An increase in disclosures was observed when MID levies were introduced. Additionally, the increase in the disclosure themes favoured by companies can be considered as a good sign. However, disclosures about climate change becoming more prevalent only in recent years show that companies listed on the SEM are late in recognising a phenomenon that has been causing changes in the Earth's climate system for at least a few decades.

This study provides new insights into environment and biodiversity reporting in the Mauritian context. It is in line with Gaia and Jones (2017) research recommendations, which state that future studies might obtain a more comprehensive portrait by extending the analysis to additional countries. Indeed, our results being close to theirs allow us to infer that similar to developed countries such as the UK, biodiversity reporting is mostly based on instrumental values. Our results also support the widespread observation of environmental reporting being mostly declarative (e.g., Ahmad & Haraf, 2013; Kuasirikun & Sherer, 2004; Nazli Nik Ahmad & Sulaiman, 2004; Thompson & Zakaria, 2004), containing little quantifiable data; general, ad hoc and self-laudatory in nature (Nazli Nik Ahmad & Sulaiman, 2004). This is a significant addition to the literature as similar studies tend to target mostly developed countries, putting developing economies such as SIDS aside even though such states are highly dependent on their environment for their economic progress.

Our findings have implications for practitioners and regulators. Regulators can use the findings of this study to issue guidelines for a better quality of disclosures. For instance, our study shows that the majority of disclosures are mostly declarative, though a change has been noticed in recent years. Climate change is a serious issue for a SIDS like Mauritius. Linked to climate change is carbon footprint. Companies have not given these two aspects due consideration. Companies should be encouraged or mandated to disclose these aspects. Voluntary initiatives by practitioners and guidelines/regulations by regulators will undoubtedly help promote practices to reduce the carbon footprint of companies and thus reduce the impact of climate change.

This study is not free from limitations. It has been carried out in a specific institutional setting, and therefore generalisation should be made with care. Furthermore, the study uses content analysis involving human effort and is prone to human error, though we took necessary steps to minimise the impact of such a mistake. Future research can extend our study sample to non-listed companies to show a comprehensive picture of biodiversity reporting in Mauritius. We relied on a single source to collect our data. Admittedly, websites are common for companies to report information. Future research can collect data from various sources to strengthen the validity of similar study using secondary data.

Acknowledgements The University of Mauritius for funding this research.
Mr. Oren Mooneeapen and Ms. Ourvashi Panchoo for research assistance for this project.

References

Adler, J. M., Dunlop, W. L., Fivush, R., Lilgendahl, J. P., Lodi-Smith, J., McAdams, D. P., McLean, K. C., Pasupathi, M., & Syed, M. (2017). Research methods for studying narrative identity: A primer. *Social Psychological and Personality Science, 8*(5), 519–527.

Ahmad, N. N. N., & Haraf, A. S. A. (2013). Environmental disclosures of Malaysian property development companies: Towards legitimacy or accountability? *Social Responsibility Journal, 9*(2), 241–258.

Air Mauritius Ltd. (2016). https://africanfinancials.com/

Alipour, M., Ghanbari, M., Jamshidinavid, B., & Taherabadi, A. (2019). Does board independence moderate the relationship between environmental disclosure quality and performance? Evidence from static and dynamic panel data. *Corporate Governance: The International Journal of Business in Society, 19*(3), 580–610.

Ane, P. (2012). An assessment of the quality of environmental information disclosure of corporation in China. *Systems Engineering Procedia, 5*, 420–426.

Baalouch, F., Ayadi, S. D., & Hussainey, K. (2019). A study of the determinants of environmental disclosure quality: Evidence from French listed companies. *Journal of Management and Governance, 23*, 939–971.

Barako, D. G., & Brown, A. M. (2008). Corporate social reporting and board representation: Evidence from the Kenyan banking sector. *Journal of Management & Governance, 12*(4), 309.

Botosan, C. A., & Plumlee, M. A. (2002). A re-examination of disclosure level and the expected cost of equity capital. *Journal of Accounting Research, 40*(1), 21–40.

Briguglio, L. (1995). Small island developing states and their economic vulnerabilities. *World development, 23*(9), 1615–1632.

Caudan Development Ltd. (2015). https://africanfinancials.com/

Chapin, F. S., Zavaleta, E. S., Eviner, V. T., Naylor, R. L., Vitousek, P. M., Reynolds, H. L., Hooper, D. U., Lavorel, S., Sala, O. E., Hobbie, S. E., & Mack, M. C. (2000). Consequences of changing biodiversity. *Nature, 405*, 234–242.

Cheah, I., & Phau, I. (2011). Attitudes towards environmentally friendly products: The influence of ecoliteracy, interpersonal influence and value orientation. *Marketing Intelligence & Planning, 29*(5), 452–472.

Cho, C. H., Roberts, R. W., & Patten, D. M. (2010). The language of US corporate environmental disclosure. *Accounting, Organizations and Society, 35*(4), 431–443.

Chu, C. I., Chatterjee, B., & Brown, A. (2013). The current status of greenhouse gas reporting by Chinese companies: A test of legitimacy theory. *Managerial Auditing Journal, 28*(2), 114–139.

Clarkson, M. E. (1995). A stakeholder framework for analysing and evaluating corporate social performance. *Academy of Management Review, 20*(1), 92–117.

Connelly, J., & Smith, G. (2003). *Politics and the environment: From theory to practice*. Routledge.
Cormier, D., & Gordon, I. M. (2001). An examination of social and environmental reporting strategies. *Accounting, Auditing & Accountability Journal, 14*(5), 587–617.
Cormier, D., & Magnan, M. (2003). Environmental reporting management: A continental European perspective. *Journal of Accounting and Public Policy, 22*(1), 43–62.
Cormier, D., Magnan, M., & Van Velthoven, B. (2005). Environmental disclosure quality in large German companies: Economic incentives, public pressures or institutional conditions? *European Accounting Review, 14*(1), 3–39.
Courtland, M. (2010). *Corporate social responsibility in hotels: How CSR programs are developed and implemented.*
Cullen, L., & Christopher, T. (2002). Governance disclosures and firm characteristics of listed Australian mining companies. *International Journal of Business Studies, 10*(1), 22–37.
Deegan, C. (2002). Introduction: The legitimising effect of social and environmental disclosures—A theoretical foundation. *Accounting, Auditing & Accountability Journal, 15*(3), 282–311.
DiMaggio, P. J., & Powell, W. W. (1983). The iron cage revisited: Institutional isomorphism and collective rationality in organizational fields. *American Sociological Review, 48*(2), 147–160.
Eade, C. (2014). *Pure white sand, clear turquoise waters and warm locals: The 10 most stunning islands in the world that we would all love to visit*. https://www.dailymail.co.uk/travel/travel_news/article-2740819/Conde-Nast-Traveller-reveals-10-stunning-islands-world.html. Accessed 3 July 2021.
Eltaib, E. E. (2012). *Environmental accounting disclosures of Australian oil and gas companies*. University of Wollongong. https://ro.uow.edu.au/theses/3777/. Accessed 3 July 2021.
ENL Land Ltd. (2010). https://africanfinancials.com/
ENL Land Ltd. (2012). https://africanfinancials.com/
Fairclough, N. (1992). Discourse and text: Linguistic and intertextual analysis within discourse analysis. *Discourse & society, 3*(2), 193–217.
Fatima, A. H., Abdullah, N., & Sulaiman, M. (2015). Environmental disclosure quality: Examining the impact of the stock exchange of Malaysia's listing requirements. *Social Responsibility Journal, 11*(4), 904–922.
Freeman, R. E. (1984). *Strategic management: A stakeholder approach*. Cambridge University Press.
Gaia, S., & John Jones, M. (2017). U.K. local councils reporting of biodiversity values: A stakeholder perspective. *Accounting, Auditing & Accountability Journal, 30*(7), 1614–1638.
Gerged, A. M. (2021). Factors affecting corporate environmental disclosure in emerging markets: The role of corporate governance structures. *Business Strategy and the Environment, 30*(1), 609–629.
Giacomini, D., Zola, P., Paredi, D., & Mazzoleni, M. (2020). Environmental disclosure and stakeholder engagement via social media: State of the art and potential in public utilities. *Corporate Social Responsibility and Environmental Management, 27*(4), 1552–1564.
Gray, R., Kouhy, R., & Lavers, S. (1995). Corporate social and environmental reporting: A review of the literature and a longitudinal study of UK disclosure. *Accounting, Auditing & Accountability Journal, 8*(2), 47–77.
Gray, R. H., Bebbington, K. J., & Walters, D. (1993). *Accounting for the environment: The greening of accountancy Part II* (Paul Chapman).
Hassan, O. A., & Romilly, P. (2018). Relations between corporate economic performance, environmental disclosure and greenhouse gas emissions: New insights. *Business Strategy and the Environment, 27*(7), 893–909.
Heras-Saizarbitoria, I., Arana, G., & Boiral, O. (2015). Exploring the dissemination of environmental certifications in high and low polluting industries. *Journal of Cleaner Production, 89*, 50–58.
Houdet, J., & Germaneau, C. (2014). Accounting for biodiversity and ecosystem services from an EMA perspective: Towards a standardised biodiversity footprint methodology. In *Accounting for biodiversity* (pp. 80–98). Routledge.

IBL Ltd. (2016). https://africanfinancials.com/

Islam, A. M., & Deegan, C. (2008). Motivations for an organisation within a developing country to report social responsibility information: Evidence from Bangladesh. *Accounting, Auditing & Accountability Journal, 21*(6), 850–874.

Jones, M. J. (2003). Accounting for biodiversity: Operationalising environmental accounting. *Accounting, Auditing & Accountability Journal, 16*(5), 762–789.

Justus, J., Colyvan, M., Regan, H., & Maguire, L. (2009). Buying into conservation: Intrinsic versus instrumental value. *Trends in Ecology & Evolution, 24*(4), 187–191.

Kabir, H., & Akinnusi, D. M. (2012). Corporate social and environmental accounting information reporting practices in Swaziland. *Social Responsibility Journal, 8*(2), 156–173.

Kamal, Y., & Deegan, C. (2011). The disclosure of corporate social and environmental-related governance practices: Evidence from the textile and garments industry of Bangladesh. In *2011 Accounting & finance association of Australia and New Zealand* (pp. 1–34). Accounting and Finance Association of Australia and New Zealand.

Kiliç, M. (2016). Online corporate social responsibility (CSR) disclosure in the banking industry: Evidence from Turkey. *International Journal of Bank Marketing, 34*(4), 550–569.

Kuasirikun, N., & Sherer, M. (2004). Corporate social accounting disclosure in Thailand. *Accounting, Auditing & Accountability Journal, 17*(4), 629–660.

Lux Island Resorts Ltd. (2013). https://africanfinancials.com/

Mahadeo, J.D., Oogarah-Hanuman, V., & Soobaroyen, T. (2011, September). Changes in social and environmental reporting practices in an emerging economy (2004–2007): Exploring the relevance of stakeholder and legitimacy theories. In *Accounting forum* (Vol. 35, No. 3, pp. 158–175). Taylor & Francis.

MCB Group Limited. (2014). https://africanfinancials.com/

Medallon, M. C., & Gallardo, M. (2014). Environmental awareness campaign: The change it brings. *Asia Pacific Journal of Multidisciplinary Research, 2*(1), 1–1.

Milne, M. J., & Adler, R. W. (1999). Exploring the reliability of social and environmental disclosures content analysis. *Accounting, Auditing & Accountability Journal, 12*(2), 237–256.

Milne, M. J., & Grubnic, S. (2011). Climate change accounting research: Keeping it interesting and different. *Accounting, Auditing & Accountability Journal, 24*(8), 948–977.

National Committee on Corporate Governance. (2004, April). *The report on corporate governance for Mauritius* (1st ed. Rev.). Port Louis: Government of Mauritius. http://www.nccg.mu/multimedia/others/editor/files/Code_of_Corporate_Governance-April%202004.pdf

National Report of the Republic of Mauritius. (2015). *Third International Conference on Small Island Developing States.* https://sustainabledevelopment.un.org/content/documents/1109215Mauritius%20National%20Report.pdf. Accessed 3 July 2021.

Nazli Nik Ahmad, N., & Sulaiman, M. (2004). Environment disclosure in Malaysia annual reports: A legitimacy theory perspective. *International journal of commerce and management, 14*(1), 44–58.

Patten, D. M. (2002a). Media exposure, public policy pressure, and environmental disclosure: An examination of the impact of tri data availability. *Accounting Forum, 26*(2), 152–171.

Patten, D. M. (2002b). The relation between environmental performance and environmental disclosure: A research note. *Accounting, Organisations and Society, 27*(8), 763–773.

Radu, C., & Francoeur, C. (2017). Does innovation drive environmental disclosure? A new insight into sustainable development. *Business Strategy and the Environment, 26*(7), 893–911.

Rahaman, A. S., Lawrence, S., & Roper, J. (2004). Social and environmental reporting at VRA: Institutionalised legitimacy or legitimacy crisis? *Critical Perspectives on Accounting, 15*(1), 35–36.

Ramdhony, D. (2017). *The influence of corporate governance practices on corporate social responsibility (CSR) reporting-evidence from Mauritius.* University of Southern Queensland. https://eprints.usq.edu.au/34620/. Accessed 3 July 2021.

Rashid, A. (2015). The influence of stakeholder power on corporate social responsibility: Evidence from a relationship-based economy. *Social Responsibility Journal, 11*(2), 270–289.

Rogers & Co Ltd. (2015). Available at https://africanfinancials.com/

Schmidt, N., & Meyer, M. D. (2009). Incorporating climate change considerations into transportation planning. *Transportation Research Record, 2119*(1), 66–73.

Statistics Mauritius. (2018). *Monthly tourist arrival*. https://statsmauritius.govmu.org/Pages/Statistics/Monthly/Monthly-Tourists.aspx

Stokes, D. L., Hanson, M. F., Oaks, D. D., Straub, J. E., & Ponio, A. V. (2010). Local land-use planning to conserve biodiversity: Planners' perspectives on what works. *Conservation Biology, 24*(2), 450–460.

Sun, D., Zeng, S., Chen, H., Meng, X., & Jin, Z. (2019). Monitoring effect of transparency: How does government environmental disclosure facilitate corporate environmentalism? *Business Strategy and the Environment, 28*(8), 1594–1607.

Sun Resorts Limited. (2015). *Annual Report*. https://africanfinancials.com/

Sylvan, R., & Bennett, D. (1994). *The greening of ethics: From anthropocentrism to deep-green theory*. University of Arizona Press.

Thompson, P., & Zakaria, Z. (2004). Corporate social responsibility reporting in Malaysia. *Journal of Corporate Citizenship, 13*, 125–136.

Ullmann, A. A. (1985). Data in search of a theory: A critical examination of the relationships among social performance, social disclosure, and economic performance of U.S. firms. *Academy of management review, 10*(3), 540–557.

Van der Laan Smith, J., Adhikari, A., & Tondkar, R. H. (2005). Exploring differences in social disclosures internationally: A stakeholder perspective. *Journal of Accounting and Public Policy, 24*(2), 123–151.

Chapter 6
Contribution to the Implementation of SDG 4: The Case of Université des Mascareignes, Mauritius

Vimi Neeroo Lockmun-Bissessur, Martin Samy, and Swaleha Peeroo

Abstract Agenda 2030 and the Sustainable Development Goals (SDGs) of the United Nations were established in order to create a more sustainable future for everyone and to address the numerous global issues the world is presently facing. Whilst higher education has historically placed more emphasis on general education and research, the diversity of its role and function has increasingly come to the fore, including service to society and the global economy. The intrinsic potential and accountability of Higher Education Institutions (HEIs) in the global SDG movement must be acknowledged, as well as their impact on the SDGs.

This paper explores the implementation of SDG 4 (Quality Education) within Higher Education, using Université des Mascareignes (UdM), a young public HEI in Mauritius, with a long-standing North–South partnership, as a case study.

A qualitative approach was used, with semi-structured interviews held with Top Management and staff of the University. A thematic analysis of the data was carried out to identify the strategies used for accomplishing SDG 4 in multiple functional areas within the institution.

Research demonstrates that, despite being a young institution with a relatively small number of students, UdM is actively addressing SDG 4, with the North–South collaboration aiding in its implementation. The advantages of the collaboration's mutually beneficial aspects, as well as its limitations are outlined.

Analysis shows that SDG 4 is addressed at UdM through the following initiatives: engagement with the community, improving access for people with special needs, improved gender representation across Faculties, offering part-time and short courses to widen student access, flexible payment arrangements for those in need, customized learning arrangements for national athletes, international scholarships for excelling students, and systematic set-up of suitable policies, procedures and guidelines in different areas. Quality Education is also ensured by addressing key components

V. N. Lockmun-Bissessur (✉) · S. Peeroo
Université des Mascareignes, Beau Bassin-Rose Hill, Mauritius

M. Samy
Kedge Business School, Talence, France

D. Crowther and S. Seifi (eds.), *Preparing for a Sustainable Future*,
Approaches to Global Sustainability, Markets, and Governance,
https://doi.org/10.1007/978-981-99-2456-1_6

such as provision of more balanced student life, ensuring a safe and healthy environment for all, through continuous student feedback mechanisms, regional and international mobility for staff and students, systematic addressing of regulatory requirements and standards, implementation of several research projects in different areas, introduction of a sustainability module across all undergraduate programs, set up of a cross-functional Quality department and additional resources for facilitating technology-enabled teaching and learning. Different means are used to train learners in sustainable development, entrepreneurship skills and employment after their studies.

This paper represents a contribution to SDG research from the developing world in Sub-Saharan Africa, an often under-represented area. It explores the impact of a long-standing North–South collaborative partnership with a young institution in Mauritius, and outlines its contribution to the sustainability agenda, thus providing insights into the multiple means by which SDG 4 can be addressed by HEIs.

Keywords Sustainable Development Goals (SDGs) · SDG 4 · Higher Education · Sustainable development · Quality Education · North–South collaboration

6.1 Introduction

The United Nations has embarked on a mission to implement the 17 Sustainable Development Goals (SDGs) through the 2030 agenda for Sustainable Development, an action plan for People, Planet and Prosperity (United Nations, 2020). This is an unprecedented global movement in order to achieve a better and more sustainable future for all and to overcome the multiple challenges that the world currently faces, related to issues including poverty, inequality, climate change, education, environmental aspects, peace and justice. Whilst Higher Education traditionally accentuated more on General Education and Research, recent development lays more emphasis on its diverse role, including service to society and the economic world at large. Promoting civic society and engaging in sustainable human development are also expected of HEIs (Shek et al., 2017). It is therefore important to recognize the inherent potential and responsibility of HEIs in the global SDG movement (Edwards & Ashida, 2020) and evaluate their contribution to the SDGs.

Although education is included as one goal within the SDGs, SDG 4 underpins the other development targets (Association of Commonwealth Universities, 2015). Education helps to eradicate poverty, to enable successful employment and decent income levels to individuals and to improve productivity that, in turn, boosts economic development. Education is furthermore a catalyst for citizen engagement, gender equality and female empowerment and promotes inclusion of disabled people. Through their educational, research and operational functions, HEIs are also at the forefront of multiple initiatives to improve health, nutrition and well-being and for the containment and eradication of diseases, for enabling greater resilience in disasters and for helping victims of crisis and conflict rebuild their lives (Australia/Pacific SDSN, 2017; Clemens & Rammel, 2015; UNESCO, 2016).

In an overview of research activities carried out internationally with regards to SDGs, the Institute for Scientific Information report revealed that SDG research is scarce within the African continent, with countries like Ethiopia, Nepal, Ghana, Tanzania, Kenya and Uganda accounting for between 0.5 and 0.9% of papers (Nakamura, 2019).

Recent research also reveals that although there is an increasing incorporation of the SDGs into business strategies, this is reviewed and communicated by only a few organizations (Junior et al., 2019). Research and evaluation cultures are important to learn lessons from the implementation of strategies, policies and actions and the Education 2030-Incheon Declaration Framework for Action for the implementation of SDG 4 highlights the importance of countries to evaluate the effect of their education policies on the achievement of SDG 4 targets (UNESCO, 2016). Nonetheless, research on frameworks for implementing the SDGs in HEIs is scarce (Filho et al., 2021).

The purpose of this paper therefore is to explore the implementation of SDG 4 within the Higher Education sector, hence contributing to research on the UN SDGs—especially SDG 4 in a sub-Saharan context, through the experience of UdM. The latter, the youngest public HEI in Mauritius with a long-standing instance of North-South collaboration, is used as an example to illustrate the ways in which SDG 4 can be acted upon in the sector.

The research question addressed is 'How and why is SDG 4 implemented within a Mauritian public HEI working in close collaboration with a European (French) university?'

6.2 Literature Review

6.2.1 Overview of SDGs

The SDGs, which constitute a shared global vision for peace and prosperity for the planet and its people, (United Nations Global Compact, 2020), were incepted in September 2015, at the UN Sustainable Development Summit in New York. Seventeen (17) SDGs were marked as global targets as the culmination of decades of work by countries and the United Nations (Allen et al., 2019).

In the context of an increasingly fragmented world, the SDGs hold an exceptional capacity to bring the world together through their worldwide applicability, with due consideration to the different national realities and specificities (Delli Paoli & Addeo, 2019). The 17 SDGs cover the social, environmental and economic scope of Sustainable Development through 169 targets and 232 unique indicators (Allen et al., 2019; United Nations Department of Economic & Social Affairs, 2020). Given the integrated nature of the SDG targets, progress towards one target is linked through complex feedbacks to other targets (Allen et al., 2019).

6.2.2 Overview of SDG4

SDG 4 is to 'Ensure inclusive and equitable quality education and promote lifelong learning opportunities for all' and aims to ensure global literacy and numeracy, equitable and quality primary, secondary and tertiary education and effective learning outcomes, to increase the number of youth and adults who have relevant skills, including technical and vocational skills (Delli Paoli & Addeo, 2019; United Nations, 2020). The focus has broadened beyond the context of primary, secondary and tertiary education to lifelong learning for all (Cheng, 2017).

Seven main targets make up SDG 4's quality and equity goals for different educational phases, whether it be vocational, technical and tertiary education (IAEG-SDGs, 2016). The first three targets stipulate that all children and adults should have access to high-quality education from early childhood through primary and secondary school as well as technical, vocational and tertiary levels. Target 4.4 strives to improve work-related and entrepreneurship skills for youth and adults. The distribution of educational access across a variety of demographic categories, with consideration to the needs of individuals with disabilities, indigenous peoples, and vulnerable groups, is the focus of Target 4.5. Target 4.6 is to ensure literacy and numeracy and significantly reduce illiteracy for youth and adults. Target 4.7 addresses the subject matter of education seeking to foster knowledge and abilities for sustainable development and sustainable lifestyles including human rights, gender equality, global citizenship and cultures of peace and nonviolence. Three additional targets (4.a, 4.b, and 4.c) are developed as means of implementation for the quality and equality aims. Target 4.a seeks to construct secure, inclusive, and productive learning facilities and environments for every child irrespective of age, gender or disability. The goal of Target 4.b is to increase the number of higher education scholarships that are accessible to developing nations, African nations and small island developing states, with the aim to improve access to higher education. Target 4.c is to boost the supply of qualified teachers through international partnerships, with consideration to least developed countries and small island developing states (IAEG-SDGs, 2016).

6.3 Higher Education and SDG 4

Higher Education has undergone rapid expansion worldwide, with student enrolment reaching 221 million in 2017, which is more than double the figure two decades ago. This expansion has prompted the emergence of a diverse HEI sector, with a multitude of HEIs with a varied range of study programs, delivery modes and student profiles (Martin & Godonoga, 2020).

In the 2015 World Education Forum in Incheon, the Incheon Declaration for Education 2030, which maps the vision for education till 2030, was adopted by the members, which included representatives from governments, multilateral and bilateral organizations and civil society, the teaching profession, youngsters as well as the private sector (UNESCO, 2016). The Education 2030 Incheon Declaration and Framework for Action acknowledges the key role of HEIs in realizing SDG 4 and accentuates on the role of HEIs to enable access to and provide lifelong learning opportunities for each and every one.

Quality education, through the teaching, learning and research activities at undergraduate, graduate, professional or executive levels, is viewed as an active contributor to sustainable development efforts at an individual or societal level and can therefore enhance capacity to support, promote and tackle the SDGs (Mori Junior et al., 2019; UNESCO, 2016). The crucial role of education as a key enabler in development and achievement of the other SDGs cannot be understated, with the need for it to be an integral part of the strategies to achieve all of the goals (Association of Commonwealth Universities, 2015; Australia/Pacific SDSN, 2017). This is a must for allowing the emergence of inclusive societies, a trained workforce and responsible citizenship (Martin & Godonoga, 2020).

HEIs therefore should be perceived as key vehicles for exploring, testing, developing and communicating conditions for transformative change for sustainable development and for contributing to the implementation of the SDGs (Clemens & Rammel, 2015; Junior et al., 2019).

Different approaches can be used by HEIs to facilitate the implementation of SDG 4. Internationalization initiatives, capacity building, research activities, the incorporation of global citizenship values, digitalization and the use of technology and flexible learning pathways are a few of the enablers, which are all elaborated upon below.

6.4 SDG 4 and Internationalization

HEIs are one of the 'primary spaces where knowledge is preserved, perpetuated and discovered', hence creating a natural overlap in the function of Higher Education and the SDGs, given the determinant role that researchers, experts and students play in endorsing ideas, products or systems relevant to the SDGs (Edwards & Ashida, 2020, p.10). HEIs operate on both the domestic and the international front, characterized by university–industry collaboration and international partnerships in Education and Research.

SDGs are built on the principle of connectedness amongst researchers, students, governments and policy-makers at various levels and in different parts of the world. HEIs can therefore enable staff and student global connectivity for internationalization of their educational and research programs through educational cooperation, collaborative research, sharing of learning and experiences and capacity building (Aarts et al., 2020). This in turn helps to inform policy decisions to address transnational challenges (Association of Commonwealth Universities, 2015).

Within Africa, internationalization has taken shape through student mobility and university partnerships within and beyond Africa, as well as transformations in university management, regional university associations, regional networks and the emerging of regionally embedded centres of excellence (Aarts et al., 2020).

Articulated and flexible education systems have also emerged, articulation symbolizing the mobility of students and learners between institutions, programs and levels of studies. This has enabled equity in education by reducing barriers and allowing for flexible transfer opportunities for students who may otherwise not be able to benefit from it (Martin & Godonoga, 2020).

6.5 SDG 4 and Capacity Building

In addition to the contribution of the internationalization process which contributes to quality education, capacity building is a key requirement that will not only help on the implementation of SDG 4 within Agenda 2030 but also several other SDGs.

SDG 4 lays emphasis on the critical mass of trained personnel across different fields, and the need to increase the supply of trained qualified teachers to least developed, developing and small island states, including international cooperation (Aarts et al., 2020).

The pandemic has catalysed the digital transformation process across the world, including Africa. Digital economies imply increased opportunities ranging from innovation, job creation, increased productivity and entrepreneurship, with a potential to contribute $180 billion to Africa's economy by 2025, representing 5.2% of its Gross Domestic Product (GDP) ('World Economic Forum', 2021). Given the digital skills and digital infrastructure gaps, this brings to the fore the need for capacity building in digital skills. This is especially important given the fact that Africa, the world's youngest population, will house almost one-fifth of the global labour force and almost one-third of the global youth labour force by 2030 ('World Economic Forum', 2021). HEIs have a determinant role to play in addressing these skills gaps and in helping to reshape education and training for the future. The COVID-19 pandemic has however heightened social disparities and made it challenging to stabilize the current socioeconomic situation, a moving objective (ADEA, 2020). This necessitates reconsidering how to provide high-quality education in a fair and inclusive manner, as well as the role of technology as one of the primary enablers (ADEA, 2020). There also needs to be a concerted effort in strengthening capacity building at the country level to enable efficient tracking of SDG indicators at all levels within Agenda 2030, given the comprehensive nature of the SDGs which require more data from a wide range of sources including administrative, financial, census, survey and learning assessment data (GUNI, 2017). Within the global education goal itself, the statistical capacity of countries is put to the test through its breadth, depth and ambition (GUNI, 2017; UIS, 2018).

6.6 SDG 4—The Key Role of Research

Besides the need for building capacity and teaching skills for employment, HEIs have a crucial role in fostering critical and creative thinking as well as producing and sharing information for social, cultural, ecological, and economic development, in addition to teaching work skills (UNESCO, 2016).

Research undertaken in strategic areas by HEIs can endeavour to find solutions to key problems across all SDGs and to contribute substantially to the creation of knowledge and enhance the creative and analytical skills required to find solutions to local and global matters regarding sustainable development (UNESCO, 2016). This can become a sounding board to inform sound policy decisions and enable national governments and donors to better meet national and international challenges through an evidence-based policy (Association of Commonwealth Universities, 2015; UNESCO, 2016), including aspects relating to education at all levels, thereby contributing to enhancing the quality of education.

The SDG 4 indicators also put a lot of emphasis on gender equality and the need to reduce inequalities. As the research community works towards the goals, it is also important that all countries participate and that authors from different backgrounds figure in scholarly publishing and journals, so as to curb inherent biases and enable the depictions of a complete and global picture of the challenges that are faced (Clarivate, 2022). In this context, explicit guidelines for funding programs to promote participation across gender or ethnic groups are increasingly common (Clarivate, 2022).

6.7 SDG 4—The Need for Global Citizenship and Community Outreach

In addition to fostering critical, analytical and creative thinking of its students and staff, Dr. Mohamedbhai, former president of the Association of African Universities, stated that 'HEIs…have a responsibility to ensure that students are sensitized to myriad development challenges, and that graduates not only find meaningful employment according to their skills, but are also equipped to create an informed and engaged citizenry' (Association of Commonwealth Universities, 2015. p. 2). Sustainable development issues, including community engagement, should hence be intrinsic to learning and teaching within institutions and mainstreamed into the curricula in order to produce leaders with the ability to realize the 2030 Agenda for sustainable development (Association of Commonwealth Universities, 2015; GUNI, 2017). Social responsibility and community engagement in HEIs can be put in place by imparting values of public service and active citizenship which would also support, nourish, and amplify the natural idealism of youth, the future leaders of tomorrow.

6.8 SDG 4 in a Post-COVID Era: Digitalization and Technology Enabled Learning

Along with the necessity of imparting values to the citizens of tomorrow, HEIs also need to use distance learning and exploit digital technologies for easing access to quality education and training (Association of Commonwealth Universities, 2015; Cheng, 2017).

The COVID pandemic has greatly accelerated and demonstrated the benefits of digital transformation and digitization throughout the world, including in Africa. The digital transformation has increased communication and partnerships between regional and international partners without the need to travel, enabled open and distance learning amongst a wider audience and eased access to quality education in different parts of the world (Lockmun-Bissessur et al., 2021).

Digital transformation can indeed contribute substantially to Africa's economic growth if the digital skills and infrastructure gaps are successfully addressed ('World Economic Forum', 2021). HEIs have the potential to play a determinant role therein.

However, the COVID-19 pandemic has also brought to light the inequalities and fragilities of society. There is a significant need for capacity building, funding and support of a strong framework for the management of crises like COVID-19 and for the long-term direction of distant education in Africa (ADEA, 2020).

6.9 SDG 4—The Need for Flexible Learning Pathways

Beyond the need for digitalization, the international Education 2030 agenda and SDG 4 emphasize on the need for education systems to adapt to flexible learning pathways, that allow entry and re-entry to all students at all ages and different educational levels, thereby making education more equitable and inclusive (Association of Commonwealth Universities, 2015; Martin & Godonoga, 2020). Education can act as a catalyst for inclusive societies if learner diversity is seen as a challenge instead of a problem and the right conditions are created to identify and develop individual talents and aptitudes in all shapes and forms (Antoninis et al., 2020).

This needs to be supported with stronger linkages between formal and non-formal learning as well as systems for the recognition, validation and accreditation of competences acquired formally and informally (Martin & Godonoga, 2020). It is nonetheless not a straightforward process that necessitates an enabling environment and well-designed implementation mechanisms with commitment at the national, institutional and individual level.

The process required is that of Recognition of Prior Learning (RPL) which is concerned with the identification, documentation, assessment and certification of learning outcomes acquired through formal, non-formal and informal means. RPL enables social inclusion and equity as the acquisition of formal qualifications allows for the employability, mobility and lifelong learning of individuals (International Training Centre of the ILO, 2022).

In summary, there are challenges in achieving SDG 4 owing to the COVID-19 pandemic and differences within the cultural, regional and international backgrounds. This brief pilot study highlights the compounding issues and considers the implementation of SDG 4 in a Mauritian HEI.

6.10 Methodology

In order to address the research question, qualitative research was carried out, which involved an interpretive and naturalistic approach. This enables one to step beyond the known and discover novel insights through participants' perspectives in their natural settings (Njie & Asimiran, 2014). It helps to further understand the phenomenon and its diversity in its natural condition and enables discoveries that contribute to the development of empirical knowledge (Corbin & Strauss, 2015; Green & Peloza, 2011; Hidayati, 2011). Mason (2002) succinctly summarises the insight obtained through qualitative research through this statement: 'Qualitative research has an unrivalled capacity to constitute compelling arguments about how things work in particular context. Through qualitative research we can explore a wide array of dimensions of the social world, including the texture and weave of everyday life, the understandings, experiences and imaginings of our research participants, the ways that social processes, institutions, discourses or relationships work and the significance of the meanings that they generate'.

The author, currently in the management team of UdM, undertook this piece of research, after ethical clearance, as part of a reflective approach for an in-depth understanding of how UdM is contributing to the implementation of SDG 4 within the working environment. It was undertaken as a pilot study in the context of her Ph.D. which centres around SDG 4 and is meant as a springboard for fine tuning the parameters for the research. Given the need for in-depth insight into the ways in which SDG 4 is addressed, interviews were identified as the most appropriate data collection technique (Creswell et al., 2007). Additional information was also obtained through websites and other relevant documentation including press media reports and newsletters. Similar qualitative methods have been used by researchers studying social responsibility or sustainability and its implementation (Green & Peloza, 2011). The research was carried out using a case study approach. A case study is a phenomenon of some sort which occurs in a bounded context (Miles & Huberman, 2014). Using a case study allows one to identify the characteristics and unique attributes of an entity, through a detailed description of a phenomenon anchored in real-life scenarios (Njie & Asimiran, 2014). In this particular research, its choice was driven by the need to thrust deep into a specific institution with specific characteristics (in this case a young public HEI in Mauritius with a long history of North–South collaboration) for a greater understanding, which would not have been possible through other means (Njie & Asimiran, 2014).

To contextualize the case, an understanding of the tertiary landscape in Mauritius is of essence. Since the country's independence in 1968, tertiary education has seen significant change in Mauritius, when there was only the University of Mauritius. Currently, there are around sixty-five (65) institutions in the tertiary sector, both public and private, with nine public HEIs under the purview of the Higher Education Commission in Mauritius (Higher Education Commission, 2022). A significant development was brought about in tertiary education in January 2019, when the Government of Mauritius introduced free higher education in public Tertiary Education Institutions (TEIs). This applies to courses leading to a first certificate, a first diploma or a first undergraduate degree only.

The Université des Mascareignes (UdM) was incepted in 2012 in Mauritius. It provides education at both undergraduate and postgraduate levels in Sustainable Development and Engineering, Information and Communication Technology and Business and Management. It is the fourth and the youngest public University in Mauritius, with a close working collaboration with the Université de Limoges, France. The bulk of the degrees are dual certificates: one from Université de Mascareignes and another from the Université de Limoges in France, according to the European LMD (Licence-Master-Doctorate) system. The University's vision is to work towards becoming a model of excellence in Higher Education, with the fulfilment of its responsibility to its territorial environment whilst maintaining a broad international outlook. Sustainability has always been on the agenda of the University, which is one of two Public HEIs in Mauritius which houses a Faculty dedicated to Sustainable Development.

Themes were identified from a critical literature review in order to draft questions for semi-structured interviews. A total of nine in-depth interviews were conducted with management and key staff of UdM until saturation was reached. These were recorded and transcribed verbatim. Each transcript was reviewed individually to identify emerging or common themes. The transcribed interviews were then coded on NVIVO software. A comprehensive thematic analysis was then undertaken, resulting in the categorization of distinct themes and sub-themes as conveyed by the participants.

6.11 Findings and Discussion

The literature review above demonstrates the inherent potential of HEIs to contribute to the SDGs. The Higher Education sector is, however, complex with a variability in vision and mission, programs, delivery modes and collaborative arrangements (Ferguson & Roofe, 2020). Given the specificities and unique attributes of each institution, the UdM is used as a case study to illustrate the different ways in which contribution is made to SDG 4, in a sub-Saharan context with a long-standing history of North–South collaboration.

6.12 Quality Education—From the Perspective of Participants

Quality Education, as defined by participants, pertains to the provision of education at a certain level within a conducive, balanced, safe and healthy environment and through adequate systems and resources—infrastructural, physical, human or technological. This should enable students to learn and thrive within the environment so as to develop into holistic human beings. Several participants emphasized on the fact that although currently more emphasis is placed on the quantitative aspects of education, the qualitative aspects of education were primordial. Participants also felt that quality should be considered in the context of sustainability. Students should be encouraged to acquire not only know-how but also to learn about 'who they are, 'how to be' and how 'to relate with others'. Education targets should therefore not be solely quantitative but should also measure those qualitative aspects, which are often disregarded.

Interviewees emphasized on the need for quality to be inherent within the system, and not be about the acquisition of labels used mainly for marketing purposes. Adherence to international standards was strongly advocated in order to ensure that the University provides a world-class quality education that can enable the mobility of students and staff.

The difference between schooling and education was also outlined by one participant—whereby it was felt that the predominant aspect of educational institutions in Mauritius is to provide schooling as opposed to providing a holistic education. The participant also mentioned about the broader social, economic and political dimension of education, which is imparted to an individual even beyond the confines of an educational institution.

6.13 SDG 4—Implementation at UdM

Findings show that UdM is assisting in the implementation of SDG4 through a number of key efforts, a few of which are outlined in this paper.

6.14 Curriculum Development with Stakeholder Involvement and Industrial Placements

One important approach enabling the achievement of SDG target 4.4 is the involvement of several stakeholders in the development of programs that are of relevance to the island, with collaborators ranging from academia and industry and beyond national borders.

Interviewees elaborated on programs that are set up in consortiums across regional and international borders. Five Masters have been set up in the different faculties through consultation with a consortium of universities as well as through the contribution of academics, subject experts and professionals through regional and international cooperation (including North–South Collaboration). Dual training programs, constituting of a mix of practical on-the-job training and classroom studies, are set up according to the needs of the industry, allowing students to work and study simultaneously, with possibilities of employment thereafter.

Joint cross-border reflection takes place in order to design programs that are better geared to the needs of the country—including potential employers and students. As an example, the Master ESPOIR was done with the collaboration of a regional health consultant, health professionals, doctors and different universities and people from Ministries—which leads to an enriched discussion through the multiple experiences of the collaborators.

Participants emphasized the importance of practicals and Industrial attachments, which are a core and mandatory part of a student's program of study at UdM from the very first year of their undergraduate study. This allows the student to gain an exposure and integrate the world of work in their field of specialization, and contribute to their job-readiness and entrepreneurial spirits. This is exemplified by the statement of a participant: '*I personally know of students who have worked on this practical hands-on module in the workshops and have successfully ventured as entrepreneurs—and this module would have greatly helped in building their confidence towards it*'.

Projects encouraging entrepreneurship in the field of sustainable development are further promoted through existing research projects including the Smart and Sustainable Campus (SSC) project, the Sustainability and Climate Change project (SCCP) and the Formation et enseignement supérieur pour la transition énergétique dans les territoires insulaires et en indianocéanie) FESTII project.

Senior Management also outlined the fact that UdM has also been designated as one of the pilot institutions for the Agence universitaire de la Francophonie (AUF) program 'Entreprendre en Afrique'. This is a three-year project to be deployed in eight countries in the sub-Saharan Africa, the Indian Ocean and the Caribbean region. It aims to foster an entrepreneurial culture within HEIs through student mentoring via a national student-entrepreneur status (SNEE) and pre-incubation centres in the AUF's Francophone Employability Centres (CEF). This is done in partnership with stakeholders in the entrepreneurial ecosystem in each of the eight hosting countries whilst strengthening innovation and research in entrepreneurship.

It was also highlighted that, with the introduction of the Higher Education Act, all programs of the UdM are to go through the accreditation process. In that light, programs of the University are being reviewed so as to ensure all the requirements are met. In a subsequent review, the programs will be reviewed jointly with the Université de Limoges (UniLim) so that they become 'Diplomes d'état' (national degrees as per the French legislation). Engineering programs are also being reviewed to ensure they meet the international Washington Accord standards. This will further enhance the international recognition of UdM degrees.

6.15 Internationalization Through North–South Collaboration

Findings indicate that Internationalization through North–South Collaboration has improved the quality of education at UdM. Through various sources of European funds (ERASMUS, MIC (International Credit Mobility) and VIF (volontaires internationals de la francophonie)), students and staff have benefited from an enriching international experience through international mobility undertaken between UdM and UniLim.

Interviewees also outlined the benefits from exchanges and visits of both academic and non-academic staff of UniLim (including the 'référents pédagogiques' (pedagogical advisors)), as well as visiting professors and researchers from other universities across the different faculties. It was stated that through the organization of training programs, workshops, seminars and conferences, the visiting professors and researchers impart their know-how for the Continuous Professional Development of UdM staff. The expertise of qualified lecturers from other (national and international) institutions is sought in order to service modules where UdM does not have qualified know-how, as for specific modules in the Master programs.

6.16 Sustainability Issues and the Need for More Balanced Student Life for Holistic Education

According to the participants, UdM has taken active measures to facilitate the implementation of target 4.7. Issues pertaining to sustainability are addressed in modules within different programs of different faculties. Additionally, two specific Masters, namely the 'Master en Efficacite Energetique et Developpement durable' and the Master in 'Sustainable Business Management' are built around the premises of sustainable development which are addressed in its modules.

One recent initiative that has been pioneered at UdM is the set-up of a 'social responsibility and environmental' module (coined as the module 'Responsable' ('Responsible')) that has been included in all undergraduate programs. The objective of the module is to inculcate a sense of initiative and responsibility amongst students and mould students into holistic global citizens of tomorrow through a hands-on, three-tiered approach throughout their undergraduate degree.

Management of UdM stated that a 'coordinator for communication and student life' has been recruited in order to enhance the student experience through a more holistic and balanced student life. In that regard, a number of cultural and musical activities are set up for the benefit of students.

6.17 Flexible Learning Pathways and Lifelong Learning

This study shows that UdM achieves targets 4.3, 4.4 and 4.6 through dispensation of programs and courses at different levels. Flexible learning pathways are provided through provision of different programs, including enrolment in diplomas, top-up programs and part-time courses. This gives opportunities for lifelong learning, whereby students get the opportunity to upgrade their skills and knowledge through different entry points whilst earning formal qualifications. Short customized courses are also offered to people according to the needs of the industry, and facilitate upskilling and reskilling.

Interviewees highlighted the flexibility provided by UdM to national sportsmen and sportswomen to attend their programs over a longer duration (five years instead of three years) so that they can successfully conciliate sports and studies.

6.18 Link Between Research and Quality of Education

Several participants highlighted the inherent link between research and quality of education at an HEI and its contribution to targets 4.3 and 4.c.

Statements of participants varied from 'a good researcher makes a good teacher' to 'Someone who does not do research will not be able to ensure quality in teaching but will also not be able to contribute optimally to projects at UdM' and 'One of the best ways to become a good teacher is to do research, as knowledge becomes obsolete over time'.

This is further exemplified through this particular statement of one participant: 'I have encouraged and supervised several staff members to obtain their PhD degree so that they can feed their teaching; which what is commonly known as research-informed teaching', as staff remain abreast with the latest advances and knowledge within their area of expertise.

Management has therefore implemented measures to encourage staff to enrol in Ph.D. programs and incentives have been provided by waiving tuition fees for staff enrolling in Ph.D.s with UdM. Furthermore, staff already holding doctorates are encouraged to mentor future researchers, and this has a direct impact on the quality of teaching.

Research is currently being undertaken across multiple fields, with many in areas of direct relevance to the Mauritian economy and context, such as refrigeration through the use of environmental-friendly carbon dioxide instead of ozone-depleting refrigerants, renewable energy, sustainability, national health concerns and so on.

Additionally, there are research projects that encourage interdisciplinary collaboration. One such example is the SCCP Program. It is a research-based program that deals with existing complex real-world problems using a systems-based and transdisciplinary approach, where all faculties are invited to contribute. The SCCP has also innovated by being inclusive through the set-up of different platforms and working groups and by encouraging dialogue at national levels, where different stakeholders are invited to participate in discussions. This includes members from within UdM and other academic institutions, from the public and private sectors, NGOs, as well as individuals with an interest to contribute to the initiative.

The SCCP program also aims to build capacity within different institutions, by encouraging enrolling of research students currently working in the public or private sector or in specialized NGOs. This is done in order to help solve the national problems and challenges by enhancing the learning and policy interface, which is deficient at the moment, as exemplified by this participant who stated that: 'There isn't a good connection between the learning that happens in tertiary and other institutions and how it connects to solving the problems and the challenges that the country faces.'

Another example is the SSC project, which aims to enable the University to adopt a fully integrated approach in educational developments, through the set-up of a living laboratory at the Rose-Hill campus of the UdM, in line with the Government's vision to transform Mauritius into a Smart Digital Island.

The international Erasmus project FESTII an AUF-COI (Agence Universitaire Francophone—Commission de l'Océan Indien) is another initiative that brings together six programs and partner universities in a North–South collaboration. The aim is to enhance capacity building in the Indian Ocean through the setup and implementation of training programs, shared research networks in the field and sensitization of stakeholders to the subject of renewable energy. Students are also mentored for initiating start-ups in renewable energy.

The importance of multidisciplinary research was emphasized whereby it will help not only in addressing the issue of a critical mass of researchers but also in finding opportunities through innovative ideas that stem from it. Such has been the case through two different projects at UdM, the SSC and the SCCP.

6.19 Quality Deployment

Along with the aforementioned actions, the commitment and support of leadership of the University to the implementation of quality initiatives were outlined as an enabler in helping to achieve the vision. Participants identified the appointment of a Head of Quality and Head of E-learning as instrumental in quality deployment.

The Quality department has contributed to the setup and implementation of several procedures and guidelines in order to address not only regulatory requirements in academia but also across different functional areas.

Program review is also undertaken internally so as to ensure programs meet the requisite standards. Joint work is undertaken with UniLim for mounting of appropriate modalities in program setup and in teaching and learning.

The Head of Quality also pilots joint work with the Independent Commission against Corruption (ICAC) through the institutionalization of the ICAC Public Sector Anti-corruption framework, which promotes a holistic approach to detect, prevent and combat corruption.

Student feedback is regularly solicited through surveys—the results of which are used for continuous improvement and enhancement of teaching quality. Several of the statutory boards and committees also include student representatives so that the student voice can be taken on board during deliberations.

The Head of e-learning is also working on the institutionalization of comprehensive policy and guidelines for technology-enabled learning, with the support of the Higher Education Commission and Commonwealth of Learning.

6.20 Teaching and Learning—Qualified Teaching Staff

Implementation of SDG 4 necessitates qualified teaching staff at all levels, addressed through target 4.c. This is being addressed through several means at the University.

New lecturers recruited are predominantly Ph.D. holders, although a Master's qualification is adequate. A few participants also outlined the importance of looking beyond qualifications and considering competence and experience along with the minimum qualifications for a post.

Furthermore, recruitment is carried out well beyond national boundaries—with international professors joining UdM whether on contractual assignments or as adjunct professors, thereby adding to the pool of experience and know-how.

Staff are also encouraged to enrol in MPhil/Ph.D. programs by waiving off tuition fees for UdM staff on Ph.D. enrolment at the University. Lecturers who are Ph.D. holders are also encouraged to mentor future researchers.

Funding has been made available in order to ensure training is continually carried out through several means and in different areas for staff, both academic and non-academic. One such recent example initiated through the Head of E-learning is pedagogy training for e-learning, where staff develop competence for online teaching and learning through an international co-diploma done in partnership with the University of Caen, Normandie, France.

Management of UdM indicated that staff from different departments and faculties have benefitted from and continue to benefit from participation in conferences and training through different means. This can be through in-house training, in-company training or training at national or international levels.

In an effort to help staff in CPD (Continuous Professional Development), they are encouraged to join professional bodies (like ITE, IMECHE, IOP and IEEE). This allows them to be in tune with the latest technology via magazines, seminars and webinars.

6.21 Occupational Safety and Health

Participants also pointed out the fact that quality of Education is indissociable from a safe and healthy environment, which is pertinent to target 4.a. At UdM, several initiatives are undertaken to promote occupational health and safety which includes amongst others, regular testing and inspection of lab equipment prior to use by students to ensure equipment are in safe working conditions, first aid training for staff and students, organization of talks, seminars and workshops on safety and health issues. A psychologist has also been appointed and can be consulted by staff and students as per need.

6.22 Equitable, Inclusive Education and Community Outreach at UdM

Besides the above-listed initiatives promoting SDG4, UdM also endeavours to enable equity and inclusiveness in education through different means, thereby partially addressing targets 4.3, 4.5 and 4.b.

Payment facilities are offered to paying students in need, by offering the option of paying in instalments. UdM also has an equal opportunity policy, which advocates for indiscriminate treatment of students and employees irrespective of age, sex, race, religion, colour, ethnic origin, nationality, disability, impairment, marital status/civil partnership status, sexual orientation, HIV status and pregnancy/maternity.

The flexible learning pathways outlined by participants above attest to the efforts towards inclusion by the University.

Female representation varies considerably across the Faculties, with a strongest representation of women in social sciences and humanities within the Faculty of Business and Management. Despite an increasing number of female students in Engineering programs, there is scope for further improving the representation of the female gender there.

The UdM also has clear guidelines for dealing with situations of conflicts of interest, thereby preventing different parties from engaging in activities or decisions that may otherwise lead to unfair decisions.

The best five (Mauritian) students from the Faculty of Sustainable Development and Engineering are now offered scholarships in order to pursue their studies overseas at the Engineering School ENSIL-ENSCI of UniLim, thereby giving an unmatched opportunity to students who would otherwise not be able to afford it. This is done through a tri-lateral agreement between UdM, the French Embassy in Mauritius and ENSIL-ENSCI. International students, if ranked amongst the best, can exercise the option for a seat secured at ENSIL-ENSCI.

Equitable education is additionally addressed through the use of appropriate pedagogy. One such concrete example is a lecturer in the Faculty of Sustainable Development and Engineering who investigated the effectiveness of the constructivist approach so as to improve the teaching and learning styles in Engineering education. This also enabled effective achievement of learning outcomes for different levels of students. In that context, 4 educational games were invented for promoting Constructivism in Engineering Education.

Interviewees also outlined that the University is working on community outreach in different ways. Flea markets have been held whereby members of the community have been invited into the University compounds for cheap purchase of goods donated by staff and students. Open days have been held in conjunction with non-Governmental organizations for free medical consultations for immediate members of the community. Along with local partners, the University is in the process of setting up a school of music for children and youngsters of different ages within the surrounding community.

Participants however mentioned the existence of a digital divide, where online sessions became a challenge for a few students. For some, it was not always straightforward to have a dedicated laptop (especially in large families with many children) or optimal internet connectivity for synchronous sessions or an appropriate dedicated workspace to adequately follow online, where background noise physically hampered the ability to follow or participate online. This became a barrier in equity of education when dispensed online. To this end, it was recommended that the University work out solutions for these students like the provision of a dedicated workspace for online access.

It was however pointed out that capital investment and infrastructural improvement are required on campus in order to promote equity and inclusiveness. Although a few efforts have been made through ramp access and dedicated lavatories, access to the disabled could be substantially improved for unhampered mobility for example through the provision of lifts. Further investment could be made to adequately equip the university library. Participants pointed out that despite the introduction of free higher education in public TEIs, that facility is not necessarily reaching the needy. Two participants advocated for a review of the scheme so as to render it more equitable, with payment from those who can afford it. This would also enable the fees collected by the University to be ploughed back into capital and infrastructural improvement.

6.23 Discussion of Findings

The tendency to work in silos and develop programs and conduct research and teaching that is occasionally divorced from the realities and needs of communities and societies is one of the challenges that is frequently cited for the Higher Education sector. UdM however is joining the group of HEIs which is internationalizing and opening up to different stakeholders across disciplines within the University, nationally, regionally and internationally in order to enrich the programs delivered, to inbuild capacity through training and research initiatives, project collaborations and community outreach activities.

Findings above substantiate statement 43 of the Education 2030 Incheon Declaration Framework for Action for the implementation of SDG 4, where it is stated that a well-established, properly regulated tertiary education system can improve quality, access, equity and relevance and can close the knowledge gap between what is taught at tertiary education institutions, including universities, and what economies and societies require. A review of the free tertiary education scheme to even out access to the especially needy appears to be of the essence and would further help the financial independence of HEIs.

Results above demonstrate that UdM is undertaking similar efforts as in certain universities worldwide (Ferguson & Roofe, 2020; Leal Filho et al., 2019; Lozano, 2006; Lozano et al., 2019), where there frequently is the development of specific programs or incorporation of subject matters relating to sustainability and sustainable development within existing programs. The introduction of a stand-alone hands-on practical 'social responsibility and environmental' module common to all undergraduate programs across the HEI however is a pioneering step which appears to be specific to UdM.

The COVID-19 pandemic has heightened the incidence of anxiety and depression amongst learners—but is often overlooked by educational institutions, as only 20 per cent of nations worldwide undertook action to offer more mental health and psychosocial support for students (United Nations, 2022). The proactiveness of UdM is demonstrated through the above findings.

Gender disparity issues are also pointed out in enrolment and success completion rates in other universities, although the nature and extent of disparity remain to be investigated from country to country and depending on the nature of programs (Ferguson & Roofe, 2020).

The inequalities in access to remote learning persist amongst different socioeconomic groups as demonstrated by the United Nations SDG Report 2022 (United Nations, 2022) and is one aspect that needs to be addressed, a point which is confirmed in the Mauritian context through the above results. Even though scholarships are partially addressing equity in access to deserving students in many universities, there remains scope for increasing the number of scholarships to those who merit in all the different Faculties and schools of UdM.

6.24 Conclusion

HEIs have a determining role to play in the achievement of the SDGs. The aim of this case study is to provide insights into ways in which implementation of SDG 4 is carried out in a young public University in Mauritius, an example of long-term North–South collaboration with a French University.

The in-depth interviews undertaken within this research demonstrate that SDG 4 targets can be addressed through a multitude of ways in different functional areas within the HEI. Findings indicate that the North–South UdM-Unilim collaboration has further catalysed the efforts of the youngest public HEI towards the achievement of several SDG 4 targets.

Program and curriculum development has been enhanced to gear them to the needs of the country. This has been done through multilateral consultations with the involvement of several stakeholders, with collaborators from academia and industry, topic specialists and professionals both within and beyond national borders through regional and international networks. The job-readiness and entrepreneurial spirit of students is facilitated through the hands-on practicals and industrial attachments, a core part of curricula within UdM. Quality of education has improved through the internationalization process, where international exchanges at different levels have sharpened the expertise of staff and enhanced the student experience through an international exposure. Sustainability issues are addressed through special programs set up in different Faculties or through specific modules included within existing ones. One recent pioneer effort in that respect is a 'Social responsibility and environmental module' that has been included in all undergraduate programs of all Faculties. Lifelong learning is also encouraged through the dispensation of customized short courses and the provision of part-time programs. Flexible study pathways are also provided for national sportsmen/sportswomen. Diverse community outreach activities at the University also contribute to the richness of staff and students' experience. The study has also demonstrated the key role of multidisciplinary research and research projects in the provision of quality education. The institutional commitment to quality, anti-corruption and occupational safety and health deployment initiatives are demonstrated as key enablers in the provision of quality education. Efforts to promote the continuous professional development of staff at all levels and for the provision of equitable education through payment facilities and scholarships offered to best-performing students have also contributed to the achievement of SDG 4 targets.

A constructive approach is important in order to enable the achievement of the SDGs worldwide, as it has been demonstrated that the pace of improvement is substantially slower than required in order to enable the realization of the targets. The COVID-19 pandemic has further highlighted and intensified the inequality and the fragility of our societies.

As the target implementation year of 2030 grows closer, and the need for higher education rises globally, this paper is intended to provide insights through the concrete examples and actions for facilitating realization of SDG 4, within a specific context of North–South collaboration and partnerships. The research indicates that much is and can be done to realize SDG 4, within the confines of the youngest public HEI in Mauritius. It is however important to involve the different stakeholders of the HEI for a more holistic, interdisciplinary approach.

This research remains however an indicative and reflective contribution and depicts the reality of one single HEI, with the limitation of being a case study that describes a certain space–time localized process.

Future research in this area could extend to exploring the challenges and opportunities regarding implementation of SDG 4 within HEIs.

References

Aarts, H., et al. (2020). *The SDGs and African higher education* (pp. 231–241). Springer International Publishing. https://doi.org/10.1007/978-3-030-14857-7_22

ADEA. (2020, August). *Impact of COVID-19 on Africa's education.* Association for Development of Education in Africa.

Allen, C., Metternicht, G., & Wiedmann, T. (2019). Prioritising SDG targets: Assessing baselines, gaps and interlinkages. *Sustainability Science, 14*(2), 421–438. https://doi.org/10.1007/s11625-018-0596-8

Antoninis, M., et al. (2020). All means all: An introduction to the 2020 Global Education Monitoring Report on inclusion. *Prospects, 49*(3–4), 103–109. https://doi.org/10.1007/s11125-020-09505-x

Association of Commonwealth Universities. (2015). *Progress and potential higher education playing its part in the Sustainable Development Goals.* https://beyond2015.acu.ac.uk/q-cards

Australia/Pacific SDSN. (2017). *Getting started with the SDGS in universities: A guide to for universities, higher education institutions and the academic sector* (Australia, New Zealand and Pacific Edition, p. 50). Sustainable Development Solutions Network. https://www.miteco.gob.es/es/ceneam/recursos/pag-web/empezar-ods-universidades.aspx

Cheng, R. (2017, August). *Getting practical about SDG 4 in teacher education in Africa*, pp. 1–6.

Clarivate. (2022). *Understanding ethnic diversity in US science.* https://discover.clarivate.com/ethnic-diversity-STEM-US-webinar?utm_campaign=EM_1_Ethnic_Diversity_LeadGen_Webinar_May18_SAR_Global_2022

Clemens, M., & Rammel, C. (2015). Brief for global sustainable development report 2015 transforming higher education for sustainable development. *Brief GSDR* (November 2014), 1–4. https://sustainabledevelopment.un.org/content/documents/621564-Mader_Rammel_Transforming%20Higher%20Education%20for%20Sustainable%20Development.pdf.

Corbin, J., & Strauss, A. (2015). *Basics of qualitative research.*

Creswell, J. W., et al. (2007). Qualitative research designs: Selection and implementation. *The Counseling Psychologist, 35*(2), 236–264. https://doi.org/10.1177/0011000006287390

Delli Paoli, A., & Addeo, F. (2019). Assessing SDGs: A methodology to measure sustainability. *Athens Journal of Social Sciences, 6*(3), 229–250. https://doi.org/10.30958/ajss.6-3-4

Edwards, S., & Ashida, A. (2020). Higher education in Japan: Internationalization, the Sustainable Development Goals and survivability. *International Journal of Comparative Education and Development.* https://doi.org/10.1108/IJCED-09-2020-0061

Ferguson, T., & Roofe, C. G. (2020). SDG 4 in higher education: Challenges and opportunities. *International Journal of Sustainability in Higher Education, 21*(5), 959–975. https://doi.org/10.1108/IJSHE-12-2019-0353

Filho, W., et al. (2019). Sustainable Development Goals and sustainability teaching at universities: Falling behind or getting ahead of the pack? *Journal of Cleaner Production, 232*, 285–294. https://doi.org/10.1016/j.jclepro.2019.05.309

Filho, W., et al. (2021). A framework for the implementation of the Sustainable Development Goals in university programmes. *Journal of Cleaner Production, 299*.https://doi.org/10.1016/j.jclepro.2021.126915

Green, T., & Peloza, J. (2011). How does corporate social responsibility create value for consumers? *Journal of Consumer Marketing, 28*(1), 48–56. https://doi.org/10.1108/07363761111101949

GUNI. (2017). *Higher education in the world—Towards a socially responsible university: Balancing the global with the local.* http://www.guninetwork.org/files/download_full_report.pdf

Hidayati, N. D. (2011). Pattern of corporate social responsibility programs: A case study. *Social Responsibility Journal, 7*(1), 104–117. https://doi.org/10.1108/17471111111114576

Higher Education Commission. (2022). *Higher Education Commission, 2022.* https://www.hec.mu/home

IAEG-SDGs. (2016). *Report of the inter-agency and expert group on Sustainable Development Goal indicators (E/CN.3/2016/2/Rev.1), BMC international health and human rights.* https://doi.org/10.1186/1472-698x-9-s1-s2

International Training Centre of the ILO. (2022). *2022 RPL recognition of prior learning, MOOC on recognition of prior learning*. https://www.itcilo.org/courses/mooc-recognition-prior-learning
Junior, R., Fien, J., & Horne, R. (2019). Implementing the UN SDGs in universities: Challenges, opportunities, and lessons learned. *Sustainability, 12*(2), 129–133. https://doi.org/10.1089/sus.2019.0004
Lockmun-Bissessur, V. N., Pistor, P., & Somanah, R. (2021). *Digitalisation of higher education institutions in Africa—Paving the way towards the attainment of the SDGs*. In Paper presented at the SARUA Colloquia Series, 9 June 2021: Higher Education Regional Collaboration: Opportunities in Complex Times? FH Münster—University of Applied Sciences, pp. 1–6. https://www.hb.fh-muenster.de/opus4/frontdoor/index/index/docId/13847
Lozano, R. (2006). A tool for a graphical assessment of sustainability in universities (GASU). *Journal of Cleaner Production, 14*(9–11), 963–972. https://doi.org/10.1016/j.jclepro.2005.11.041
Lozano, R., et al. (2019). Teaching sustainability in European higher education institutions: Assessing the connections between competences and pedagogical approaches. *Sustainability, 11*(6), 1–17. https://doi.org/10.3390/su11061602
Martin, M., & Godonoga, A. (2020, March). *SDG 4-policies for flexible learning pathways in higher education taking stock of good practices internationally*, p. 52. https://doi.org/10.13140/RG.2.2.31907.81449
Mason, J. (2002). *Qualitative researching*. Sage.
Miles, M., & Huberman, M. (2014). Qualitative data analysis. *CEUR Workshop Proceedings*, pp. 89–92.
Nakamura, M. (2019). Navigating the structure of Research on SDGs. *Asian Journal of Multidimensional Research (AJMR), 8*(12), 113. https://doi.org/10.5958/2278-4853.2019.00330.6
Njie, B., & Asimiran, S. (2014). Case study as a choice in qualitative methodology. *IOSR Journal of Research & Method in Education (IOSRJRME), 4*(3), 35–40. https://doi.org/10.9790/7388-04313540
Shek, D., Yuen-Tsang, A., & Ng, E. (2017). University social responsibility (USR): Insight from the historical roots to the contemporary challenges. In *University Social Responsibility and Quality of Life* (pp. 99–120). Springer. https://doi.org/10.1007/978-981-10-3877-8
UIS. (2018). *Quick guide to education indicators for SDG 4*, p. 48.
UNESCO. (2016). *Education 2030—Incheon declaration framework for action for the implementation of Sustainable Development Goal 4*, p. 83. http://unesdoc.unesco.org/images/0024/002456/245656E.pdf
United Nations. (2020). *Transforming our world—The 2030 agenda for sustainable development*. https://sdgs.un.org/2030agenda
United Nations. (2022). *The Sustainable Development Goals report 2022*. https://unstats.un.org/sdgs/report/2022/The-Sustainable-Development-Goals-Report-2022.pdf
United Nations Department of Economic and Social Affairs. (2020). *The SDGs in action*. https://www.undp.org/content/undp/en/home/sustainable-development-goals.html
United Nations Global Compact. (2020). *United Nations Global Compact 20th anniversary campaign overview*.
'World Economic Forum'. (2021). *2021 how Africa can lead an inclusive cohesive and sustainable pandemic recovery*. World Economic Forum. https://www.weforum.org/agenda/2021/01/how-africa-can-lead-an-inclusive-cohesive-and-sustainable-pandemic-recovery/

Chapter 7
The Impact of Leadership on Engagement at Work: A Comparison Between Temporary and Staff Workers

Maria João Santos and Flávio Oliveira

Abstract The study realised had, as its main objective, analysed if there are significant differences in the engagement levels prevailing among temporary workers and staff workers. Additionally, it also examined the type of contractual relationship, and if it had any influence on the perceptions of workers regarding the leadership practices on going in the organisation. In the empirical research, a survey was carried out by means of a questionnaire to 82 workers, 63.4% of which were from a temporary work company. With regards to leadership, the staff work group is the one that most refers to having transformational leaders and is the one that best evaluates the performance of their leader. In contrast, workers originating from temporary work companies consider having a higher number of "laisser-faire" leaders. In terms of engagement at work, there were no significant differences. However, temporary workers show a greater expenditure of energy, high mental resilience, willingness to invest and insist on the difficulties that arise in the work context ("vigour"). They also showed greater immersion ("absorption"), in the sense of total concentration in the work activities in which they are involved. On the other hand, effective workers show greater dedication related to a high level of involvement, performance and providing meaning to the professional activity carried out.

Keywords Engagement at Work · Transactional leadership · Transformational leadership · Temporary workers · Staff Workers · Employment contract

This work was supported by FCT, I.P., the Portuguese national funding agency for science, research and technology, under the Project UIDB/04521/2020

M. J. Santos (✉) · F. Oliveira
CSG/SOCIUS—ISEG—Universidade de Lisboa, Lisbon, Portugal

D. Crowther and S. Seifi (eds.), *Preparing for a Sustainable Future*,
Approaches to Global Sustainability, Markets, and Governance,
https://doi.org/10.1007/978-981-99-2456-1_7

7.1 Introduction

To obtain greater flexibility, companies frequently make recourse to the utilisation of human resources supplied by temporary employment agencies. Temporary work assumes a triangular relationship involving (i) the temporary employment agency (that contracts, pays and exercises disciplinary power over the worker); (ii) the company contracting the temporary employment agency (that hosts the worker on their installations and exercises powers of authority and management) and (iii) the temporary worker. This triangular relationship enables companies to manage their labour requirements more flexibly given they incur no legal obligation as regards the employment contracts and may therefore more easily integrate or eliminate the working position in accordance with their own needs. This practice, as an atypical employment relationship, frequently associated with the existence of high levels of precariousness and uncertainty, whether of an economic, sociologic or psychological nature, among the workers involved, is highly regulated (in keeping with the legal regime of each state) to ensure this practice is not used either abusively or generally rather than serving as a recourse for deploying only in exceptional situations.

From the user company/client perspective, this practice, despite the advantages resulting from great labour flexibility, incurs a series of problems. There is frequent reference to how temporary workers encounter greater difficulties in adapting to the organisational culture and establishing trusting relationships with the employer as they understand from the outset that their contractual situations are only temporary. Other authors also highlight the difficulties in exercising their functions as these workers do not always gain access to the same sources of information as their peers, full-time members of staff. In addition, as they normally perform their functions for limited periods throughout the weekly timetable, they experience greater difficulties in establishing strong connections with either their colleagues or their own spaces and correspondingly often do not get involved as deeply in activities as members of staff with long-term contractual relationships.

Despite the literature identifying some problems, there is a need for deeper analysis of whether there are significant differences in the engagement levels prevailing among temporary workers and staff workers. Does the type of contractual relationship influence the perceptions of workers as regards the leadership practices ongoing in the organisation? Does this also shape the levels and characteristics of engagement? Furthermore, are there any differences in the perceptions as regards organisational leadership practices and the aforementioned levels of engagement? Across these facets, just what differences might exist between staff on full-time, permanent contracts and workers sourced from temporary employment agencies?

Based on these initial questions, we carried out the present study with the core objectives of (a) identifying the leadership practices deployed in organisations; (b) verifying to what extent workers engage with their jobs; (c) verifying the existence of statistically significant differences between temporary workers and staff workers and (d) ascertaining whether the type of contractual relationship (full-time employees with permanent contracts and workers from temporary employment agencies) influences the perceptions of these workers towards both the organisational leadership practices and engagement. Despite the theoretical evidence about the relationship between leadership styles and job engagement, there is little known about whether these variables correlate and, above all, their interactions across these two groups of workers in keeping with their different contractual relationships with the company where they work. In order to respond to these research questions and objectives, we designed the current study.

Structurally, this study contains three sections. The first includes the theoretical framework that spans the concept of leadership and, in particular, concepts of transactional and transformational leadership as well as the relationships identified between job engagement and leadership. The second part presents the methodology applied, details the data collection instruments, the procedures and the statistical processes carried out. This also includes the statistical analysis taking into account both the descriptive analysis and the inference of results. The third section sets out and discusses the results returned. The study then closes with the most relevant conclusions that detail answers to the questions above and the respective study objectives.

7.2 Literature Review

7.2.1 Transactional and Transformational Leadership

Leadership has emerged as an increasingly important area of study, especially in recent decades. Recent research in the leadership field has particularly focused on transactional leadership, which according to Bass (2009), *cit in* Noro et al. (2015) represents the utilisation of monitoring and controlling employees, correspondingly deploying rational or economic metrics and on transformational leadership (Bass & Avolio, 1993; Rowold & Rohmann, 2009; Tims et al., 2011), which Rowold (2011) describes as focusing not on obtaining immediate objectives but rather on changing and transforming the organisation and its members of staff.

7.2.1.1 Transactional Leadership

According to Bass (2009), *cit in* Noro et al. (2015), transactional leadership applies rational or economic metrics to monitor and control workers. In this leadership style, there is an exchange between the leader and the follower in which one influences the other in order for both parties to receive something of value. Burns (2003), *cit in* Noro et al. (2015), states that the exchange in transactional leadership extends beyond transactions of a merely economic type but also includes exchanges of a political and psychological nature between leaders and the led, for example, promotions in the career structure (contingency rewards). Non-financial rewards vary from the most tangible (extra holidays and days off) to the less tangible (praise and recognition). In this leadership style, the leaders provide rewards for meeting targets. However, to obtain satisfactory results, there has to be clarity over the targets and objectives alongside recognition for meeting such targets.

Transactional leadership concentrates closely on the exchanges between the leaders and the led and susceptible to application in two different fashions. One involves contingent rewards that reflect how leaders request the completion of a task and with a predetermined reward attributed to the led after completing this task. Alternatively, another approach is management by exception that only involves leaders in conveying feedback in a negative manner as a corrective measure. Management by exception incorporates the ways leaders set about observing the led and issuing negative feedback whenever they commit errors, thus, punishment for unproductive behaviours (Northouse, 2016). However, as Rowold (2011) identifies, these targets normally undergo definition based on the requirements of the organisations and not the individual characteristics of workers, requiring individualised considerations and/or adjustments to the objectives in keeping with individual preferences.

Bass and Riggio (2006) propose there are three components to contingent reward-based transactional leadership, management by exception and *laissez-faire* leadership. Thus, contingent reward is the first of three factors for transactional leadership. This displays an exchange process between leaders and workers in which the efforts of the latter are exchanged for specific rewards. With this type of leadership, leaders attempt to obtain the agreement of workers over what should be done and what the rewards shall be for those completing these tasks. As regards management by exception, this leadership involves corrective criticism, negative feedback and negative reinforcement. Management by exception takes two different forms: active and passive. Leaders that adopt the active management by exception style closely observe workers in the search for errors or infringements of the rules and, subsequently, take corrective measures. On the other hand, leaders deploying a passive style only intervene after the standards are not met and/or problems arise. *Laissez-faire* leadership refers to leaders who do not make any particular contribution.

7.2.1.2 Transformational Leadership

Transformational leadership does not seek to return immediate results but rather to change and transform the organisation and its members of staff. According to Rowold (2011), transformational leaders communicate values of a higher order than the explicit working tasks for each team member individually. The same author states that these leaders evaluate the track records, the values and motives of each team member in order to formulate a common vision for a better future. This definition implies that transformational leaders seek out an agreement within groups and develop the abilities and resources of staff to better cope with future needs.

According to McWhinney (1997), *cit in* Noro et al. (2015), transformational leadership features a rare quality—charisma—, which is "the power to capture and stimulate the follower". Kouzes and Posner (2013), *cit in* Noro et al. (2015), declare transformational leadership to be "the art of mobilising others to want to fight for shared aspirations". In turn, Fairhurst (2001), *cit in* Noro et al. (2015), defines how the main role of charismatic leaders involves applying their vision to create meanings and symbols of change for their workers.

Transformational leadership concentrates on leaders as the source of inspiration for the led to exceed their own expectations. Transformational leaders display captivating personalities and magnetism able to attract the led and fostering incentives and support that motivate the led to want to achieve the objectives set. One of the factors inherent to transformational leadership, according to the aforementioned author, is inspirational motivation described as the means by which leaders provide incentives to the led to participate and contribute to implementing the vision, convincing them that their contributions are vital and irreplaceable. An additional factor derives from intellectual stimulation that describes how these leaders motivate the led, challenging them to think "outside of the box", encouraging creative expression and supporting them in all their efforts to obtain their team objectives (Northouse, 2016).

Transformational leadership also incorporates different components. The most commonly applied classification is that put forward by Bass and Avolio (1997): (a) Idealised influence: transformational leaders behave in ways that enable them to serve as models for their employees. The leaders are admired, respected and trusted (Bass & Riggio, 2006); (b) Inspirational motivations: transformational leaders behave in ways that motivate and inspire those around them, supplying meaning and challenges to the work of their staff; (c) Intellectual stimulation: transformational leaders stimulate the efforts of their staff to become innovative and creative, questioning assumptions, reformulating problems and approaching old situations in new ways; (d) Individualised consideration. Transformational leaders pay particular attention to the needs of each individual follower for their self-realisation and growth, acting as a trainer or mentor (Bass & Riggio, 2006).

Transformational leadership constitutes one of the most prevalent paradigms in the contemporary literature on leadership and associates this with various results in employees, such as well-being, creativity and the performance of tasks. This type of leadership is, as Burns (1979) defends, the process through which leaders and their teams help each other to progress to higher levels of morale and motivation. Hence, transformational leaders are morally mature individuals able to motivate the behaviours and attitudes of workers to attain higher levels of moral rationale (Burns, 1979). However, the process of transformational leaders influencing the attitudes and behaviours of their staff is anything but simple (Avolio et al., 2009).

7.2.2 Job Engagement and Leadership

The term engagement refers to the involvement, satisfaction and enthusiasm for the job (Harter et al., 2002). Emerging out of the work of Kahn (1990), engagement describes the intimate involvement and the structure of the working experience. When engaged, employees are emotionally connected to other persons and cognitively monitor the team leadership (Harter et al., 2002). Engagement arises whenever workers know what to expect, have the resources to complete their jobs, participate in growth and feedback opportunities and feel that they make significant contributions to their organisations/companies (Batista-Taran et al., 2009).

The definition of job engagement spans the ways workers physically, cognitively and emotionally connect with their workplace functions (Kahn, 1990). Nevertheless, there are many variations on the definitions of job engagement with that most commonly adopted coming from Schaufeli and Bakker (2004) and stating that this encapsulates a positive spirit, and gratification in relation to the job which reflects in vigour, dedication and absorption. This definition spans three components: vigour—a high level of energy during work; dedication—to be deeply involved in the work; and absorption—to be concentrated and happy at work. Furthermore, the appropriateness of work, the affective commitment and the psychological climate emerge as potential antecedents of job engagement (Bakker & Albrecht, 2018). Furthermore, the literature considers the functions of leaders as one of the factors influencing job engagement (Altinay et al., 2019; Besieux et al., 2018).

In a broader context, the literature reports a relationship between leadership in general and job engagement. For example, Altinay et al. (2019) establish a significant connection between leadership and involvement at the relational level. In the same perspective, Bakker et al. (2010) propose that transformational leadership provides a catalyser for job engagement. Based on this stance, transformational leadership may represent an antecedent for job engagement in keeping with its influence over a series of resources (for example, autonomy or constructive feedback) that subsequently interrelate with engagement (Besieux et al., 2018). Transformational leadership also associates with engagement in the workplace because this seeks to transform the mentality of individuals within the framework of obtaining organisation and team targets and objectives.

Leaders influence the efficiency of their organisations through their workers. Leadership may have a major impact on the engagement of workers within their organisations. However, transactional leadership limits the scope of leaders to reward-based behaviours as the means of obtaining the best performance from their staff, which only returns short-term effects. In turn, transformational leadership alters the approaches of workers from isolated individuals to members of a greater group. When workers perceive that they are members of a collective, they tend to support the values and objectives of a group, which raises their motivation to contribute towards the greater good (Bakker & Schaufeli, 2008). The same authors also refer to how transformational leaders provide an inspiring vision of the targets that may help in overcoming their own interests and restrict the emergence of factions in organisation. They draw on new and broader energies from their staff. Bakker and Schaufeli (2008) argue that workers who interact with their leaders raise their levels of engagement. Hence, the transformational style of leadership leads to increased organisational commitment and the satisfaction of employees with their work, and leaders are thus able to concentrate on building relationships and developing trust that further deepens levels of engagement. Transformational leaders do not appear as remote figures of power but rather as mutual support for a common objective, the collective good of organisations. In this perspective, transformational leaders hold the capacity to directly influence the levels of engagement of workers and are capable of attending to the human and working needs of their staff, a dividend from a highly unique and empowering style (Nohria et al., 2008).

Positive leadership styles have interrelated with various indicators for the well-being of workers, such as their job engagement, defined as a persistent and wide-reaching affective-cognitive state that does not focus on any object in particular, any event, individual or behaviour (Schaufeli & Bakker, 2004). This is also implicit in the personal engagement that takes advantage of the individual members of organisations in their job functions. Hence, in their workplace engagement, people express themselves physically, cognitively and emotionally while performing their roles (Kahn, 1990). Based on this theory, job engagement reflects a motivational concept as workers apply their personal resources to complete their employment tasks (Christian et al., 2011).

According to Kahn (1990), job engagement means workers are (1) physically involved, (2) cognitively vigilant and (3) maintain empathic connections with other persons in the workplace. Hence, job engagement spans three dimensions: the physical, cognitive and emotional components. However, Macey and Schneider (2008) perceive engagement from a broad perspective and make the distinction between psychological state engagement, behavioural engagement and trait engagement. Irrespective, engagement emerges within the scope of leadership and both mutually and positive influence each other.

7.3 Methodology

7.3.1 Objectives and Data Collection Instruments

The general objective of this study involves analysing to what extent two groups of workers (those from temporary employment agencies and staff workers) engage in their work. In particular, this seeks to (i) verify whether there are statistically significant differences in engagement between temporary workers and staff workers; (ii) ascertain whether the type of work shapes the perceptions workers hold about the leadership practices ongoing in the organisation and, finally, (iii) identify whether there is any relationship between the perceptions of the organisation's leadership practices and the engagement of workers. It should also be noted that in the particular case of temporary workers, the study exclusively analysed their relationship with the company they actually work for without analysing the relationship they established with the staff agency, which is also significant and can obviously affect the results found.

To undertake the study, we made recourse to a non-probabilistic, accidental sample by convenience made up of 82 workers. As the data collection instrument, we deployed a questionnaire that integrated sociodemographic and professional characteristics (ad hoc, designed for this purpose), engagement, evaluated through the *Utrecht Work Engagement Scale* (UWES-17) put forward by Schaufeli and Bakker (2010), adapted for European Portuguese by Teles et al. (2017), and the Multifactor Leadership Questionnaire (MLQ, 5X), proposed by Avolio and Bass (2002), according to the Dias version (2007). The scales applied in this study are not the only options for evaluating engagement and leadership but nevertheless stood out as the most appropriate for this study.

Job engagement seeks to evaluate to what extent the staff feel involved with their work (and not the effectiveness of the work carried out) with this questionnaire structured into seventeen items divided into three interrelated dimensions: "vigour", with six items, "dedication" with five and "absorption" with six. Participants responded to each item in accordance with a Likert-type ordinal scale with seven alternatives, ranging from 0 (if you never had that feeling or belief) and 6 (that you felt or have frequently felt). This scale, in its Portuguese UWES-17 version, was submitted to a sample made up of 1,369 Portuguese social welfare workers, with the reliability of the results evaluated through Cronbach's alfa and validated through exploratory factorial analysis. The results obtained by Teles et al. (2017) are in compliance with those of earlier studies as regards to the internal consistency, both in terms of the 17 items and for the three dimensions ("vigour", "dedication" and "absorption").

The UWES-17 psychometric study returns good psychometric properties in addition to a very high overall level of Cronbach's alfa, coming in at 0.937 (cf. Table 7.1).

To evaluate the leadership type, we applied the Multifactor Leadership Questionnaire (MLQ, 5X) put forward by Avolio and Bass (2002) in accordance with the Dias version (2007). We opted for this version, without any alterations given that Dias carried out the validation of the scale for the Portuguese population in 2007 through its respective psychometric testing. The result obtained by the internal consistency determined a global Cronbach's alfa for the 45 items of 0.908, thus reporting very

Table 7.1 Internal consistency of the UWES-17 dimensions

Dimensions	Average	Dp	R/item total	R^2	α without item
Vigour	21.93	5.422	0.916	0.840	0.878
Dedication	22.22	5.613	0.855	0.760	0.920
Absorption	25.44	6.559	0.857	0.757	0.929
Global Cronbach's alfa coefficient		**0.937**			

Table 7.2 Grouping of dimensions and subdimensions

Dimension	Subdimension	Variables
Transformational leadership —**TF**	Idealised influence (attributed)—**IIA**	10,18,21,25
	Idealised influence (behavioural)	6,14,23,34
	Motivational Inspiration	9,13,26,36
	Intellectual Stimulation	2,8,30,32
	Individualised Consideration	15,19,29,31
Transactional leadership —**TA**	Contingent Reward	1,11,16,35
	Manage/Lead by active exception	4,22,24,27
	Manage/Lead by passive exception	3,12,17,20
Laissez-Faire—non-management—**LF**	—	5, 7, 28, 33
Performance of Leaders—**PF**	Extra Effort (**EEF**)	39, 42, 44
	Effectiveness (**EFF**)	37,40,43,45
	Satisfaction (**SAT**)	38,41

Source Adapted from Dias (2007)

good internal consistency, which enabled Dias (2007) to conclude that the 45 variables refer to the same dimension thereby enabling the determination of the role of leadership.

The aforementioned author grouped the variables in accordance with the key provided by the MLQ, 5X model (Avolio & Bass, 2002: 105–109) (cf. Table 7.2).

According to the Table 7.2, the variables 10, 18, 21 and 25 describe the Idealised Influence IIA subdimension. The Cronbach's alfa result for this variable came in at 0.707, which reflects how the data provide a reasonable description of the same characteristic (cf. Table 7.3). In overall terms, the variables describe their respective subdimensions to a reasonable or good extent and only return weak results for the Idealised Influence (behaviour)—IIc and Manage/Lead by passive exception—mbe (p) subdimensions. However, in these cases, the Cronbach's alfa still verge on reasonable values. In relation to the dimensions, the subdimensions provide a very good description of Transformational Leadership and the Performance of Leaders (alfa results of 0.900 and 0.904, respectively); a good description of Non-Management (alfa = 0.867), with Transactional Leadership receiving only a weak description by its respective subdimensions (alfa = 0.607).

MLQ, 5X represents a questionnaire containing 45 questions with each attributed a number that indicates the direction of the attitude of the participant in relation to each affirmation. The answers to these 45 questions adopt a Likert-type scale from 1 to 5 (0—Never, 1—Once in a while, 2—Sometimes, 3—Frequently, 4—Frequently if not always, 5—No Opinion) and, in accordance with their answers, participants characterise their leadership perceptions. A particular group of questions approache each style of leadership in a total of 45 questions making up the questionnaire as presented in Table 7.3.

Table 7.3 Internal dimensions and subdimensions

Dimension	Subdimension	Cronbach's alfa for the subdimension	Cronbach's alfa for the dimensions
Transformational leadership	Idealised influence (attributed)	0.707	0.900
	Idealised influence (behavioural)	0.698	
	Motivational Inspiration	0.874	
	Intellectual Stimulation	0.868	
	Individualised Consideration	0.764	
Transactional leadership	Contingent reward	0.778	0.607
	Manage/Lead by active exception	0.755	
	Manager/Lead by passive exception	0.688	
Laissez-Faire—non-management	—	0.867	0.867
Performance of Leaders	Extra Effort	0.839	0.904
	Effectiveness	0.871	
	Satisfaction	0.852	

Source Adapted from Dias (2007)

7.3.2 *Ethical Considerations*

We distributed the data collection instrument through *Google Docs* and sent an email to all workers requesting they respond online. This invitation contained all the instructions necessary for their participation. Any research study should follow the principles set out in the Helsinki Declaration and the Convention of Oviedo and hence the drafting of this study strove to thoroughly comply with these methodological considerations. Hence, throughout this process, we guaranteed the ethical and deontological procedures as regards the confidentiality and anonymity of participants as well as the voluntary nature of their participation alongside the opportunity to withdraw at any stage in the research. All of the information gathered is subject to anonymity and does not contain any personal identification details, thereby ensuring the confidential handling of the data. We codified the data and applied the statistical procedures following which the data shall be destroyed.

7.3.3 Statistical Procedures

Quantitative research applies statistical analysis for three core purposes: to describe the data, to test the hypotheses and to supply evidence as regards the measured properties of the quantified variables (Marôco, 2018).

Thus, following data collection, we carried out a first stage of analysis on all questionnaires in order to eliminate any that may have been left incomplete or erroneously completed. Subsequently, we advanced with the codification and tabulation in order to prepare for the statistical processing of the 82 individuals making up the sample obtained.

For statistical processing, we made recourse to inductive or inferential descriptive statistics. In relation to the descriptive statistics, we determined: the absolute and percentile frequencies, trend measures (median_M) and dispersion and variability measures (standard deviation_Dp, minimum_Min. and maximum_Máx.).

We verified the adherence to the norm for variable distribution by applying the Kolmogorov–Smirnov (K–S) test with the Lilliefors correction (given we do not know the median and standard deviation for this universe) (Pestana & Gageiro, 2020). The decision to apply non-parametric tests stems from the absence of one or more of the conditions necessary for applying parametric tests with their non-parametric alternatives correspondingly eligible for application. The present study carried out testing on the sample distribution norm as a means of supporting decisions over the statistical testing procedures subsequently applied (cf. Table 7.4). The results of the Kolmogorov–Smirnov test indicate the sample does not display a normal distribution and indeed falls far short of normal ($p < 0.05$) in almost all factors, which was then confirmed by the histograms (cf. Annex II), which report the bias of each factor, resulting in the utilisation of non-parametric tests.

Table 7.4 Results of the Kolmogorov–Smirnov test

	Kolmogorov–Smirnov			Shapiro–Wilk		
	Statistic	df	Sig	Statistic	df	Sig
Vigour	0.124	82	0.003	0.918	82	0.000
Dedication	0.163	82	0.000	0.892	82	0.000
Absorption	0.145	82	0.000	0.886	82	0.000
Total_Engagement	0.172	82	0.000	0.872	82	0.000
Transformational leadership TF	0.062	82	0.200	0.984	82	0.425
Transactional leadership TA	0.104	82	0.028	0.983	82	0.361
Laissez-Faire Leadership LF	0.154	82	0.000	0.954	82	0.005
Profile of leaders	0.064	82	0.200	0.984	82	0.382

As regards the inductive or inferential statistics, we adopted non-parametric tests in keeping with how almost all these factors fail the normality test: the Mann–Whitney U test (U), the Spearman correlation test (Ró) with the following linear association criteria: $R < 0.2$ very weak; $0.2 \leq R < 0.4$ weak; $0.4 \leq R < 0.7$ moderate; $0.7 \leq R < 0.9$ high; $0.9 \leq R < 1$ very high (Pestana & Gageiro, 2020).

The level of test significance was established at 0.05 in keeping with the convention that identifies this as the minimum alfa acceptable for scientific research (Pestana & Gageiro, 2020).

All the statistical processing took place through the SPSS (*Statistical Package For The Social Science*) 26.0® programme for WINDOWS®.

7.4 Presentation of Results

7.4.1 Descriptive Analysis

- **Sociodemographic and professional characteristics of participants**

The total sample registers a male majority (68.3%) with almost half of the workers falling into the 36–45 age group (47.6%), with 48.8% respondents Brazilian in nationality and 47.6% Portuguese while over two-thirds (67.1%) of respondents hold a higher education qualification, with the majority employed by temporary employment agencies (63.4%) and with another clear majority having been employed by the agency for a period of over one year (67.3%). Of the 30 participants who are staff members, 43.3% have held their positions for over one year. As regards their professional statuses, the respondents are, with the same percentage levels (10.97%, respectively), production assistants and social assistants, followed by directors (8.53%) (cf. Table 7.5).

- ***Engagement***

The statistics portraying engagement across the sample range from a minimum of 12 to a maximum of 94, corresponding to an average of 69.59 ± 16.63. This reports that the dimension with the highest average is absorption ($M = 25.44 \pm 6.55$), with vigour registering the lowest result ($M = 21.93 \pm 5.42$). Hence, we may state that these workers report total concentration in the employment activities in which they are engaged (cf. Table 7.6).

- **Leadership**

The statistics returned by the leadership subdimensions detail how the highest average level corresponds to transformational leadership ($M = 48.80 \pm 14.81$), trailed by transactional leadership ($M = 24.05 \pm 3.63$). We obtained a still lower average result for *Laissez-Faire* leadership, 5.73 ± 3.11. As regards the profile of leaders, this

Table 7.5 Sociodemographic and professional characteristics of participants ($N = 82$)

Variables	*N*	%
Sex		
Male	56	68.3
Female	26	31.7
Age		
18–25	11	13.4
26–35	13	15.9
36–45	39	47.6
46–55	15	18.3
> 55	4	4.9
Nationality		
Portuguese	39	47.6
Brazilian	40	48.8
Other	3	3.7
Education		
4th grade (1st cycle of primary education)	1	1.2
7th/8th/9th years (3rd cycle primary education)	4	4.9
Secondary school	22	26.8
University	55	67.1
Type of work		
Temporary	52	63.4
Staff	30	36.6
Length of time with the temporary employment agency		
Up to 1 month	1	1.9
1–3 months	3	5.8
3–6 months	3	5.8
6–12 months	10	19.2
> 1 year	35	67.3
Length of time with a permanent contract		
Up to 1 month	3	10.0
1–3 months	6	20.0
3–6 months	5	16.7
6–12 months	3	10.0
> 1 year	13	43.3

Table 7.6 Statistics on workplace engagement

Dimensions	*N*	Min	Max	M	SD
Vigour	82	6	30	21.93	5.42
Dedication		4	30	22.22	5.61
Absorption		1	34	**25.44**	6.55
Total engagement		12	94	69.59	16.63

Table 7.7 Statistics on the leadership subdimensions

Subdimensions	*N*	Min	Max	M	SD
Transformational leadership—TF	82	15	78	**48.80**	14.81
Transactional leadership—TA		14	34	24.05	3.63
Laissez-Faire—non-management—LF		0	12	5.73	3.11
Performance of Leaders—PF		1	36	22.21	7.47

gained an average of 22.21 ± 7.47. The results convey how workers perceive transformational leadership as a style of leadership in which leaders provide incentives, inspire and motivate innovation and generate changes that help shape and enhance the future success of the company. At the executive level, this may derive from building a strong sense of corporate culture. Workers experience this as a management style projecting greater space for their creativity, looking to the future and coming up with new solutions for problems (cf. Table 7.7).

7.4.2 *Inferential Analysis*

The results stemming from the inferential analysis provide the responses to the defined objectives.

- ***Relationship between the employment contract and the type of leadership***

The findings point to a statistically significant influence of transformational leadership ($p = 0.029$) and which rises among the staff workers to indicate how they are the group that pays the most attention to the type of leadership present. Equally, the profile of leaders generates a statistically significant influence ($p = 0.11$), with staff workers once again attributing the greatest importance, thus emerging as the group with the best perceptions of the performance of leaders. On the contrary, we would highlight how temporary workers return a higher average ranking for *laissez-faire* leadership (cf. Table 7.8).

- ***Relationship between the type of employment contract and engagement***

Table 7.8 Mann–Whitney U test between leadership and type of work

Type of work	**Temporary**	**Staff**	**UMW**	***p***
Leadership	**Average Ranking**	**Average Ranking**		
Transformational leadership—TF	37.13	49.07	553.000	**0.029**
Transactional leadership—TA	39.27	45.37	664.000	0.262
Laissez-Faire—LF	44.38	36.50	630.000	0.146
Performance do Leaders—PF	36.41	50.32	515.500	**0.011**

There are no statistically significant differences between engagement and the type of employment contract. Nevertheless, the average ranking results of temporary workers emphasise vigour and absorption to a greater extent as well as scoring higher on total engagement. However, staff workers return a higher ranking in dedication (cf. Table 7.9).

- ***Relationship between the type of leadership and engagement***

This reports moderate to very high correlations between the type of leadership and the dimensions of engagement with almost all not only positive but also statistically significant with the exception of transactional leadership. There are negative correlations between *Laissez-Faire* leadership and the three dimensions of engagement, which indicates that the greater the extent workers perceive the *Laissez-Faire* type of leadership, the lower is their vigour, dedication and absorption. In the remainder of the correlations, there are positive and direct associations that convey how stronger worker perceptions of the leadership as transformational and transactional bring about greater levels of vigour, dedication and absorption, hence, greater engagement in work. In relation to the performance of leaders, the correlations are positive and with statistically significant differences portraying how when workers hold more positive perspectives of the characteristics of their leaders, their vigour, dedication and absorption levels rise reflecting in greater commitment to the work (cf. Table 7.10).

Table 7.9 Mann–Whitney U test between engagement and type of work

Type of work	**Temporary**	**Staff**	**UMW**	***p***
Engagement	**Average Ranking**	**Average Ranking**		
Vigour	41.99	40.65	754.500	0.806
Dedication	39.78	44.48	690.500	0.387
Absorption	42.72	39.38	716.500	0.540
Total_Engagement	41.86	40.88	761.500	0.859

Table 7.10 Spearman's correlation results for the dimensions of engagement and the type of leadership

Engagement	Vigour	Dedication	Absorption
Type of leadership		R_{s*}	
Transformational leadership—TF	0.480	0.428	0.424
		p	
	0.000	0.000	0.000
Transactional leadership—TA		R_{s*}	
	0.091	0.040	0.096
		p	
	0.417	0.719	0.392
Laissez-Faire Leadership—LF		R_{s*}	
	−0.346	−0.399	−0.357
		p	
	0.001	0.000	0.001
Performance of Leaders—PF		R_{s*}	
	0.440	0.397	0.366
		p	
	0.000	0.000	0.001

* R_s = Representation of Spearman's coefficient; p = significance

7.4.3 Discussion of Results

Based on this sample of 82 workers, we may report that:

(1) The two groups of workers (full-time, permanent staff and those from temporary employment agencies) report high levels of engagement. The statistics detailing engagement indicate an average of 69.59 ± 16.63 for the total sample (staff and temporary workers). The dimension with the highest average value was "absorption" (M = 25.44 ± 6.55), with "vigour" registering the lowest level (M = 21.93 ± 5.42).

As referenced in the literature review, job engagement, by definition, consists of a positive mental state of well-being in relation to the working context (Teles et al., 2017). Schaufeli et al. (2002), state that engagement configures a positive affective-cognitive process, persistent and wide-reaching, interrelating with the work and made up of three dimensions: behavioural ("vigour"), emotional ("dedication") and cognitive ("absorption"). "Vigour" portrays the levels of energy, mental resilience, desire to invest and insisting despite the difficulties emerging in the workplace context; "dedication" relates to the nature of the involvement, spirit/mood, sense of pride and challenge, inspiration in performing the job and attributing significance and meaning to the professional activities undertaken; while "absorption" reflects the "immersion" and total concentration on the working activities to which they are committed, encountering difficulties in letting go of work and conveying the sensation that time passes swiftly. This absorption in work is, therefore, a state of attention and intrinsic satisfaction. Workplace engagement therefore encapsulates a significant level of identification with the work and the high levels of energy dedicated. Hence, we may state that the results returned suggest this sample of workers reports a high level of absorption, reflecting in greater workplace involvement. Absorption reflects the concentration and involvement in work, which renders the passage of time intangible, and with workers encountering difficulties in disconnecting from their duties (Hayati et al., 2014).

(2) In terms of the subdimensions of leadership, transformational leadership (M = 48.80 ± 14.81) returned the highest level trailed by transactional leadership (M = 24.05 ± 3.63) with Laissez-Faire leadership (M = 5.73 ± 3.11) returning by far the lowest level. As regards the profile of leaders, this obtained an average of 22.21 ± 7.47.

In accordance with Hayati et al. (2014), these results may stem from how transformational leadership contains a robust personal component, which enables leaders to motivate their followers, inspiring them and incorporating transformations into their attitudes and driving the achievement of objectives sustained by values and ideas. Hence, transformational leaders stand in opposition to transactional leaders, who appeal more to the individual conveniences of their followers based upon the fundamental assumption that transactional leadership represents a process related to recognising the performances obtained (Gemeda & Lee, 2020). Altinay et al. (2019) establish a significant connection between transformational leadership and higher levels of involvement in the workplace.

(3) Following analysis of the relationship between the types of employment contract (full-time staff and temporary workers) and the leadership type, we may report that staff workers rank transformational leadership higher ($p = 0.029$). This constitutes a statistically significant influence conveying how this group considers the type of leadership present in their organisation is of the transformational type. Equally, for this same group of employees, the profile of leaders holds statistically significant greater relevance ($p = 0.11$). This group of workers (staff) attributed the greatest weighting, thus, those who best evaluate and perceive the performance of their leaders.

On the contrary, temporary workers return a higher average ranking for laissez-faire leadership. According to the results obtained and taking the literature into account, we may state that transformational leadership may represent an antecedent to workplace engagement to the extent this influences a series of resources, for example, autonomy and constructive feedback. The literature interlinks transformational leadership with workplace engagement to the extent this is able to transform the mentalities of persons to obtain the targets and goals of organisations and teams (Besieux et al., 2018). These results therefore corroborate the findings of Khan et al. (2020) in which both the temporary and staff workers held stronger perceptions about the type of transformational leadership present in the companies where they worked.

(4) In studying the relationship between these two groups of workers and engagement, we did not encounter any statistically significant differences between the level of engagement and the type of contractual employment. However, the average values attributed demonstrate that temporary workers display greater "vigour", "absorption" and also rank higher for total engagement. Staff workers rank higher in terms of "dedication", with this facet potentially reflecting greater involvement in their work. Dedication shares much in common with job involvement, perceived as the level of attachment and identification with the job (Hayati et al., 2014).

(5) Following analysis of the relationship between the leadership type and engagement, we here report moderate to very high correlations between the type of leadership and the dimensions of engagement, almost all positive in direction and displaying statistically significant differences, with the exception of transactional leadership, at the level of $p > 0.05$. Furthermore, negative correlations emerge between Laissez-Faire leadership and the three dimensions of engagement, which indicates how the greater the worker perception of the Laissez-Faire leadership type, the lower the level of "vigour", "dedication" and "absorption". In the remaining correlations, there are positive direct associations conveying that the greater the worker perception of their leadership as transformational and transactional, the greater their "vigour", "dedication" and "absorption", thus, the deeper their involvement in the work. In relation to the performance of leaders, the correlations are positive and with statistically significant differences, thus suggesting that the greater the appreciation of workers for the characteristics of their leaders, the greater their "vigour", "dedication" and "absorption", reflecting in their greater involvement in their work.

The relationship between transformational leadership and job engagement has received a great deal of academic attention. Specifically, Schaufeli and Bakker (2004) refer to how engagement, as a positive state of spirit, gratifying and related to the job, is characterised by vigour, dedication and absorption. Bakker et al. (2011) argue that the direct relationship between transformational leadership and job engagement displays different intensities under different conditions. The same authors state there is no simple, direct influence and this may result from assisting employees to interpret the meaning of their work, thereby portraying how dedication encapsulates a sense of meaning, enthusiasm, inspiration, pride and challenge (Schaufeli & Bakker, 2010).

The study by Fong-Yi et al. (2020) also reports that Laissez-Faire management leads to less engagement, especially as regards vigour and, the higher the level of worker perceptions of their leadership as transformational, the greater their vigour, dedication and absorption, thus reflecting in greater workplace involvement. Hence, job engagement mediates a positive relationship between transformational leadership, job performance and helpful behaviours. Nevertheless, we need to refer to how temporary employment contracts display certain disadvantages for workers: the fact that workers do not hold great expectations in relationship to long-term employment leads to greater job insecurity as these workers not only do not know how long they may retain their current positions but also whether the temporary employment agency will provide them with another position at another company. There are also difficulties in gaining promotion in the organisations where they are then working.

7.5 Conclusion

In conclusion, as regards the central question to which this research sought to respond: the comparison between i) staff workers and ii) workers supplied by temporary employment agencies as regards their perceptions of leadership and job engagement, the findings portray the existence of differences between these two groups of workers.

As regards transformational leadership and the profile of leaders, there is a statistically significant influence with staff workers having attributed the highest ranking. Thus, this group of workers most refers to having transformational leaders and who return the best evaluations of the performance of their leaders. On the other hand, workers holding temporary contractual relationships as they arrive from temporary employment agencies, despite also reporting a high level of valuation for this variable, were those considering having higher levels of "laisser-faire" leaders. This situation may indicate the fact that leaders display differentiated attitudes and behaviours in keeping with whether they are interacting with staff or temporary workers, attributing less orientation and planning time to this latter group of workers.

As regards to job engagement, there are no statistically significant differences between each of these two groups. Nevertheless, we would highlight that the average rankings attributed display the presence of differences between these two groups. Temporary workers report greater "vigour" and "absorption" and score higher in the job engagement total. In contrast, staff workers rank higher in terms of "dedication", with this potentially reflecting greater involvement in their jobs.

Therefore, temporary workers report spending higher levels of both energy and mental resilience, greater desires to invest and persist in overcoming the difficulties that arise within the workplace context ("vigour"). This also extends to deeper levels of immersion ("absorption") in terms of their total concentration in the labour activities they undertake. However, staff workers display a greater level of "dedication" related to their high levels of involvement, enthusiasm, pride, challenge and inspiration in performing their job tasks and attributing meaning to their professional activities.

In summary, both these groups of workers display high levels of engagement even while we would stress that the ways in which this engagement emerges differs fairly significantly and requires further reflection in order to grasp the underlying reasons, especially why temporary workers express higher levels of energy and concentration in managing their workplace activities. This study also warns of the sheer importance of establishing differentiated strategies for managing workers in keeping with whether they are temporary or full-time workers.

This work only holds the intent of establishing an instrument for reflection, enabling evidence-based practices, without any definitive conclusions or drawing a line under the discussion. *In fine*, the results returned provide the opportunity to recognise the existence of new relationships, presenting new hypotheses for discussion and proposing contributions to the debate on the professional management of human resources and this study thus represents a contribution towards future studies.

This approach contains certain methodological limitations, especially stemming from having deployed a self-response questionnaire as its data collection tool given that the veracity of these results is inherently dependent on the sincerity of participant answers, hence, their answers may have been driven by considerations of socially desirable behaviours, potentially not reflecting such behaviours in practice, which may have biased the final study results. However, we would maintain that the data collection tool aligns with the purposes of the study and emerged as a line of orientation for the research questions and objectives set. Despite the sample containing a considerable number of workers, this does not enable any generalisation of these results. As such, this study requires confirmation through the undertaking of other studies in the same area, whenever possible, with larger samples that are representative of the population under study. We would also reference that the transversal study approach may constitute another methodological limitation, which raises certain advantages and disadvantages. We may consider it advantageous to carry out a briefer and methodologically less complex study. However, there are disadvantages such as the evaluation focusing on the moment of data collection, thus, on the retrospective perceptions of respondents, without any follow-up to the issues under study, a situation that might have been avoided by opting for a longitudinal study. It should also be noted that in the particular case of temporary workers, the study exclusively analysed their relationship with the company they actually work for without analysing the relationship they established with the staff agency, which is also significant and can obviously affect the results found. Another limitation relates to the scarcity of studies approaching the transformational and transactional leadership styles and their implications on the job engagement of temporary workers.

Even aware of these limitations, we believe that our methodological options were the most appropriate for this type of study and have duly enabled the production of more concrete knowledge on the phenomenon under study. As future lines of research, it would be interesting to replicate this study with a wider sample in order to compare and contrast the findings with this present study. Finally, we must emphasise that the times prevailing worldwide, due to the COVID-19 pandemic, also emerge as a major factor of limitation on this study as many temporary workers saw their contracts end and were not subject to renewal. Such a situation naturally drives reductions in the levels of engagement of these workers given that they knew their contracts would not be renewed. Thus, it would be important to replicate this study in a post-pandemic context.

References

Altinay, L., Dai, Y., Chang, J., Lee, C., Zhuang, W., & Liu, Y. (2019). How to facilitate hotel employees' work engagement: The roles of leader-member exchange, role overload and job security. *International Journal of Contemporary Hospitality Management, 31*(3), 1525–1542.

Avolio, B. J., & Bass, B. M. (2002). *Developing potential across a full range of leadership: Cases on transactional and tranformational leadership*. Psychology Press.

Avolio, B. J., Walumbwa, F. O., & Weber, T. J. (2009). Leadership: Current theories, research, and future directions. *Annual Review of Psychology, 60*(1), 421–449.

Bakker, A. B., & Demerouti, E. (2008). Towards a model of work *engagement. Career Development International, 13*(3), 209–223.

Bakker, A. B., & Albrecht, S. (2018). Work engagement: Current trends. *Career Development International, 23*(1), 4–11.

Bakker, A. B., & Schaufeli, W. B. (2008). Positive organizational behavior: Engaged employees in flourishing organizations. *Journal of Organizational Behavior, 29*, 147–154.

Bakker, A. B., & Bal, P. M. (2010). Weekly work engagement and performance: A study among starting teachers. *Journal of Occupational and Organizational Psychology, 83*(1), 189–206.

Bakker, A. B., Albrecht, S. L., & Leiter, M. P. (2010). Key questions regarding work engagement. *European Journal of Work and Organizational Psychology, 20*(1), 4–28.

Bakker, A. B., Albrecht, S. L., & Leiter, M. P. (2011). Key questions regarding work engagement. *European Journal of Work and Organizational Psychology, 20*(1), 4–28.

Barbosa, F. M., Gambi, L. D. N., & Gerolamo, M. C. (2017). Liderança e gestão da qualidade–um estudo correlacional entre estilos de liderança e princípios da gestão da qualidade. *Gestão & Produção, 24*, 438–449.

Bass, B. M. (1990). From transactional to transformational leadership: Learning to share the vision. *Organizational Dynamics, 18*(3), 19–31.

Bass, B. M., & Avolio, B. J. (1993). Transformational leadership and organizational culture. *Public Administration Quarterly, 17*(1), 112–121.

Bass, B. M., & Avolio, B. J. (1997). *Full range leadership development: Manual multifactor leadership questionnaire*. Menlo Park, Mind Garden.

Bass, B. M., & Riggio, R. E. (2006). *Transformational leadership*. Psychology Press.

Batista-Taran, L. C., Shuck, M. B., Gutierrez, C. C., & Baralt, S. (2009). The role of leadership style in employee engagement. In M. S. Plakhotnik, S. M. Nielsen, & D. M. Pane (Eds.), *Proceedings of the eighth annual college of education & GSN research conference* (pp. 15–20). Florida International University. http://coeweb.fiu.edu/research_conference/

Besieux, T., Baillien, E., Verbeke, A. L., & Euwema, M. C. (2018). What goes around comes around: The mediation of corporate social responsibility in the relationship between transformational leadership and employee engagement. *Economic and Industrial Democracy, 39*(2), 249–271.

Burns, J. M. (1979). *Leadership*. Harper & Row.

Castanheira, P., & Costa, J. A. (2011). In search of transformational leadership: A (meta)analysis focused on the Portuguese reality. *Procedia Social and Behavioral Sciences, 15*, 2012–2015.

Christian, M. S., Garza, A. S., & Slaughter, J. E. (2011). Work engagement: A quantitative review and test of its relations with task and contextual performance. *Personnel Psychology, 64*(1), 89–136.

Dias, P. A. B. (2007). *A liderança em escolas da Região Centro: identificação e avaliação de perfis de liderança: acção transformacional e influência na organização escolar*. (Dissertação

de Mestrado). Viseu: Universidade Católica Portuguesa. Centro Regional das Beiras. Acedido em https://catalogo.bibliotecas.ucp.pt/cgi-ibin/koha/opacdetail.pl?biblionumber=238288
Noro, G. B., Stüker, C., & Oliveira, J. H. R. (2015). A relação das características inerentes aos perfis de liderança no desenvolvimento do processo coaching. *Revista Capital Científico-Eletrônica* (RCCe)-ISSN 2177–4153, *13*(1), 59–76.
Downton, J. V. (1973). *Rebel leadership: Commitment and charisma in the revolutionary process.* Free Press.
Fong-Yi L., Hui-Chuan T., Szu-Chi L., Yu-Chin L. & Cheng-Chen L. (2020). Transformational leadership and job performance: The mediating role of work engagement. *Sage Open, 10*(1), 21582440198.
Fullan, M. (2002). The change leader. *Educational leadership. Beyond Instructional Leadership, 59*(8), 16–21.
Fullan, M. (2003). *Liderar numa cultura de mudança.* Edições Asa.
Gemeda, H. K., & Lee, L. (2020). Leadership styles, work engagement and outcomes among information and communications technology professionals: A cross-national study. *Heliyon, 6*(4), e03699.
Harter, J. K., Schmidt, F. L., & Hayes, T. L. (2002). Business-unit-level relationship between employee satisfaction employee engagement, and business outcomes: A meta-analysis. *Journal of Applied Psychology, 87*, 268–279.
Hayati, D., Charkhabi, M., & Naami, A. (2014). The relationship between transformational leadership and work engagement in governmental hospitals nurses: A survey study. *Springerplus, 3*, 25.
Kahn, W. A. (1990). Psychological conditions of personal *engagement* and dis*engagement* at work. *Academy of Management Journal, 33*, 692–724.
Khan, H., Rehmat, M., Butt, T. H., Farooqi, S., & Asim, J. (2020). Impact of transformational leadership on work performance, burnout and social loafing: A mediation model. *Future Business Journal, 6*(1), 1–13.
Khan, M. J., Aslam, N., & Riaz, M. N. (2012). Leadership styles as predictors of innovative work behavior. *Pakistan Journal of Social and Clinical Psychology, 9*(2), 17–22.
Macey, W. H., & Schneider, B. (2008). The meaning of employee *engagement. Industrial and Organizational Psychology, 1*, 3–30.
Marôco, J. (2018). *Análise estatística com o SPSS Statistics* (6th ed.). ReportNumber.
Nohria, N., Groysberg, B., & Lee, L. (2008). Employee motivation: A powerful new model. *Harvard Business Review, 86*, 78–84.
Northouse, P. G. (2016). *Leadership: Theory and practice. 7.ª Edição.* Sage Publications, Inc.
Pestana, M. H., & Gageiro, J. N. (2020). *Análise de dados para as Ciências Sociais—A complementaridade do SPSS*. Edições Silabo.
Rowold, J., & Rohmann, A. (2009). Transformational and transactional leadership styles, followers' positive and negative emotions, and performance in German nonprofit orchestras. *Nonprofit Management and Leadership, 20*(1), 41–59.
Rowold, J. (2011). Relationship between leadership behaviors and performance: The moderating role of a work team's level of age, gender and cultural heterogeneity. *Leadership and Organization Development Journal, 32*(6), 628–647.
Schaufeli, W. B., & Bakker, A. B. (2004). Job demands, job resources and their relationship with burnout and *engagement*: A multi-sample study. *Journal of Organizational Behavior, 25*, 293–315.
Schaufeli, W. B., & Bakker, A. B. (2010). Defining and measuring work engagement: Bringing clarity to the concept. In Bakker A. B. Leiter M. P. (Ed.), *Work engagement: A handbook of essential theory and research.* Psychology Press, pp. 10–24.
Schaufeli, W., Bakker, A., & Salanova, M. (2006). The measurement of work *engagement* with a short questionnaire: A cross-national study. *Educational and Psychological Measurement, 66*, 701–716.

Schaufeli, W. B. and Salanova, M. (2007). Work *engagement*: An emerging psychological concept and its implications for organizations. In S. W. Gilliland,D. D. Steiner. and D. P. Skarlicki (Eds.), *Research in social issues in management: Managing social and ethical issues in organizations* (vol. 5). Information Age Publishers, pp. 135–177.

Schaufeli, W. B. (2013). The measurement of work *engagement*. In R. R. Sinclair, M. Wang, L. E. Tetrick (Ed.), Research methods in occupational health psychology: Measurement, design, and data analysis. Routledge/Taylor & Francis Group, pp. 138–153.

Schaufeli, W., Martinez, I., Marques Pinto, A., Salanova, M., & Bakker, A. (2002). Burnout and engagement in university students: A cross national study. *Journal of Cross-Cultural Psychology, 33*, 464–481.

Schaufeli, W. B., Salanova, M., Gonzalez-Roma, V., & Bakker, A. B. (2002). The measurement of *engagement* and burnout: A two sample confirmatory factor analytic approach. *Journal of Happiness Studies, 3*, 71–92.

Sergiovanni, T. J. (2004) *O mundo da liderança: desenvolver culturas, práticas e responsabilidade pessoal nas escolas*. Edições ASA.

Sergiovanni, T. (2005). The virtues of leadership. *The Educational Forum, 69*, 112–123.

Teles, H., Ramalho, N., Ramalho, V., & Ribeiro, S. (2017). Adaptação e validação da Utrecht Work Engagement Scale (UWES) aplicada a assistentes sociais em Portugal. *Adaptação e validação da Utrecht Work Engagement Scale (UWES) aplicada a assistentes sociais em Portugal* (2), 10–20.

Tims, M., Bakker, A. B., & Xanthopoulou, D. (2011). Do transformational leaders enhance their followers' daily work engagement? *The Leadership Quarterly, 22*(1), 121–131.

Part III
Preparing for the Future

Chapter 8
The Sustainability Reporting in the Municipalities: A Study in Portugal

Cristina Góis, Maria Costa, and Clara Viseu

Abstract Sustainability practices are increasingly used and the target of concern by society in general. The citizens are increasingly more interested in environmental, economic, and social issues. The way in which this sustainability is promoted by public entities makes it possible to assess the extent to which society's concerns are addressed by these entities. The municipalities, which are the main objective of this study, are framed within the public sector, and they play a fundamental role in the daily life of the population, having as their main mission to implement policies that defend the interests and needs of the population, through the availability efficient and transversal public services.

Knowing the importance of the municipalities for all aspects of society and the way they influence it, concerns about reporting information on sustainability have increased in recent years, giving special attention to environmental, economic, and social factors. This work was developed with the objective of evaluating how much and what type of information on sustainability is reported in the various information disclosure documents presented by the municipalities in the north of Portugal. In addition to assessing the level of disclosure of public entities, we seek to identify the main factors that determine this disclosure.

Despite the importance of the topic under study, there are still very few works that analyze this subject. This research aims to contribute to overcoming this lack of study and also to verify the weaknesses and strengths of disclosure regarding sustainability considering the information disclosed by municipalities of the northern region of Portugal on their websites.

The present investigation analyzes and quantifies the matter of disclosure of sustainability in 86 Municipalities belonging to the North region of Portugal, using the analysis of their web pages and investigating the possible explanatory factors for their level of disclosure. After identifying the explanatory factors, namely the sociodemographic, socioeconomic, fiscal, and political factors, the disclosure index was prepared. The results obtained suggest that the disclosure of sustainability is

C. Góis (✉) · M. Costa · C. Viseu
Polytechnic Institute of Coimbra, Coimbra Business School, ISCAC, Coimbra, Portugal

D. Crowther and S. Seifi (eds.), *Preparing for a Sustainable Future*,
Approaches to Global Sustainability, Markets, and Governance,
https://doi.org/10.1007/978-981-99-2456-1_8

average, as can be seen by the result that 56.77% of the Municipalities disclose information. The category that presents the greatest disclosure is economic, followed by social, general, contracting services and public works, and, finally, environmental. The last presents a visible lack of disclosure on the web pages, thus establishing an incentive for municipalities to improve their performance systems regarding the information made available in this particular area, which is increasingly crucial.

Keywords Sustainability reporting · Municipalities · Education · Political competence · Transparency · Websites

8.1 Introduction

Sustainability practices are increasingly used and the target of concern by society in general. The citizens are increasingly more interested in environmental, economic, and social issues.

The way in which this sustainability is promoted by public entities makes it possible to assess the extent to which society's concerns are addressed by these entities. The municipalities, which are the main objective of this study, are framed within the public sector, and they play a fundamental role in the daily life of the population, having as their main mission to implement policies that defend the interests and needs of the population, through the availability efficient and transversal public services.

Knowing the importance of the municipalities for all aspects of society and the way they influence it, concerns about reporting information on sustainability have increased in recent years, giving special attention to environmental, economic, and social factors. This work was developed with the objective of evaluating how much and what type of information on sustainability is reported in the various information disclosure documents presented by the municipalities in the north of Portugal. In addition to assessing the level of disclosure of public entities, we seek to identify the main factors that determine this disclosure.

Despite the importance of the topic under study, there are still few works that analyze the subject. This research aims to contribute to filling this lack of study and also to verify the weaknesses and strengths of disclosure in terms of sustainability, considering the information disclosed by the municipalities of the northern region of Portugal on their websites.

The present investigation analyzes and quantifies the issue of disclosure of sustainability in 86 Municipalities belonging to the North region of Portugal, using the analysis of their web pages and investigating the possible factors that explain their level of disclosure. After identifying the explanatory factors, namely sociodemographic, socioeconomic, fiscal, and political factors, the disclosure index was prepared. The study begins with the collection of information on sustainability, using the analysis of web pages related to all the selected municipalities.

To achieve the stated objective, the work is divided into six chapters. After this chapter, a literature review is presented, where the topic of social responsibility is analyzed and theories associated with the dissemination of information are identified. In the third part, the main global references for the dissemination of social responsibility are presented. In the fourth chapter, the research design is defined and in the fifth part, the results obtained are analyzed and discussed. The study ends with the presentation of the main conclusions.

8.2 Literature Review

It is considered that the theme of social responsibility began to be addressed in the eighteenth century, with the publication of Adam Smith's book, "The Wealth of Nations" (1776). According to Francischini et al. (2005), the works of Charles Eliot (1906), Arthur Hakley (1907), and John Clark (1916) also addressed this issue, but their relevance was reduced.

The publication by Howard R. Bowen of the book Social Responsibilities of the Businessman (1953) is considered the foundation for the study of corporate social responsibility (CSR). According to the author, by social responsibility of businessmen, meant "the obligations of businessmen to pursue those policies, to make those decisions, or to follow those lines of action that are desirable in terms of the objectives and values of our society." Caulkins et al. (2013) consider that study was the first comprehensive discussion of business ethics and social responsibility. It created a foundation by which business executives and academics could consider the subjects as part of strategic planning and managerial decision-making. As a result, many experts (e.g. Carroll [1999]) believe it to be the seminal book on corporate social responsibility.

The concept of corporate social responsibility has a wide range of definitions, as it depends on the cultural and religious influences and/or habits of each particular society, but it is also interrelated with the concept of sustainability, in which aspects are assimilated. economic, social, and environmental (Bandeira, 2010). Responsible organizations must follow a management model based on those three aspects, which is usually called the "Triple Bottom Line." With regard to the economic aspect, organizations are concerned with making a profit, fulfilling all their obligations, and still seeking to create adequate conditions for their stakeholders. The social category is related to the impacts of organizations on the social systems in which they operate. Finally, with regard to environmental aspects, concerns are mainly focused on the impacts of organizations on ecosystems (GRI, 2015).

One of the milestones of social responsibility in Portugal was the holding of the European Summit in Lisbon in 2000 and the publication of the Green Paper (2001) on corporate social responsibility. The summit resulted in the need to provide a new impetus to Community policies, with the aim of establishing a new strategic goal for the European Union and the strengthening "(…) of employment, economic reform and social cohesion within the framework of an economy based on in knowledge" (European Parliament, 2000). The Green Paper aimed to promote a European framework for corporate social responsibility.

Currently, corporate social responsibility is considered a decisive factor in the development and growth of companies, which leads consumers to opt for products originating in companies that are not directly or indirectly involved in corruption actions, which are transparent and respect the environment and the community (Francischini et al., 2005). This orientation attracts the most qualified professionals, who choose to collaborate with organizations that value the quality of life of employees and respect their rights.

Organizations to take the initiative to disseminate information on social responsibility are usually not based on a single objective but on several. These objectives can be analyzed according to different theoretical frameworks, so the voluntary disclosure of SR information can be presented through the use of different theories, among which are the stakeholder theory, the theory of legitimacy, and the theory of the agency.

Stakeholders are people or groups that depend on an organization to achieve its objectives and goals, and on whom, in turn, the organization depends (Neto, 2019). Eugénio (2010) argues that in the application of the stakeholder theory, managers are motivated to disclose information to stakeholders with the power of influence, in order to show that they are acting in accordance with the expectations of these same stakeholders. This theory presents theoretical support that is based on the identification of the existence of a stimulus for organizations to carry out voluntary social disclosures (Deegan, 2002). It is also verified that stakeholders have increased their attention to the acts that public entities practice, examining whether they constitute economic, social, and environmentally responsible behavior, which means that they increasingly assess the sustainable behavior of these entities (Ribeiro et al., 2018).

The legitimacy is directly related to a "social contract" between the community and the entity that must be elaborated or modified according to society's approval (Slack & Shrives, 2008). For Deegan (2002), the theory of legitimacy is based, like others, on a theory of oriented systems that considers the interrelationship between companies and the society in which they operate, thus defending that entities are not only influenced but also influence. as well as the society in which they are inserted.

The theory of legitimacy is a powerful theory for understanding voluntary social and environmental disclosures, serving as the basis for several studies of the disclosure of social responsibility information (Tilling, 2004). In addition to the aforementioned author, several authors consider legitimacy as the central axis of institutional theory (Deegan, 2002; Joseph, 2010).

In the scope of the public sector, there are several studies that adopt the institutional theory alone or in combination with others, such as the theory of legitimacy, in order to justify the implementation of sustainability practices and dissemination of information in public sector entities (Joseph, 2010; Joseph & Taplin, 2011; Lodhia, 2010; Marcuccio & Steccolini, 2005 and Mussari & Monfardini, 2010). The theory of legitimacy is based on the principle that, in order to continue to operate successfully, it is necessary for companies (with an emphasis on public entities) to act within the boundaries that society recognizes as being the markers of socially and environmentally acceptable behavior, in order to carry out this identification and, in turn, that the evaluation of this information is necessarily disclosed (Ribeiro et al., 2018).

In the agency theory, proposed by Jensen and Meckling (1976), the existence of a contract is identified, which the authors define as a contract formulated between one or more people (the principal) who hire another individual (the agent) to provide a service on its behalf, where it is understood that it has some power of authority in decision-making. The agent is hired to perform a job and has the function of making decisions and carrying out actions that provide high-value benefits to the organization, maximizing profits and the value of the entity. It should be noted that the agent will not always act in the best interest of the principal. In this way, and to solve this problem, the principal can limit the divergences of its interest by establishing appropriate incentives for the agent and incurring monitoring costs aimed at limiting the agent's activities. These costs are called agency costs.

The study by Ahmad et al. (2003) identified a relationship between disclosure levels in account reports and agent monitoring costs. Thus, it is concluded that this theory is associated with the levels of disclosure of information on social responsibility and that it can cause an increase or decrease in the information disclosed.

Most studies related to social responsibility worldwide are mainly oriented towards the private sector. Sciulli (2009) shows concern about the scarcity of research in the public sector focusing on sustainability issues. However, in recent years, it is possible to identify an increase in research in the public sector where, among the three areas of social responsibility (economic, social, and environmental), where environmental aspects dominate the analysis by researchers (Burritt & Welch, 1997; Ribeiro & Guzmán, 2011, 2012; Sciulli, 2009; Tilt, 2001).

The role of the public sector in addressing social responsibility has increased in recent years. In this way, the government is encouraged to improve its responsibility and transparency in the management of public resources, since social responsibility is considered an intrinsic function of the public sector.

One of the objectives of social responsibility is defined as the practice of actions that increase the quality of life of society in general. However, despite social responsibility being present in several sectors, only the public sector considers social responsibility as its institutional practice. It is up to the public sector to carry out political and public actions that meet the needs of society such as education, security, transport, culture, housing, and others that provide social inclusion and an improvement in conditions of life (Frey et al., 2008).

In Portugal, in the wider sphere of the public sector, we find the municipalities. The formation of Portuguese Municipalities "(…) dates back to the Middle Ages and is a product of the circumstances of the reconquest, as a form of self-organization of territorially based communities. At that time, the main concern of feudal lords was war and not the management of their domains" (Bilhim, 2004).

Despite its existence since ancient times, changes were introduced by the Portuguese revolution of 1974, with the emergence of the Constitution of the Portuguese Republic, of 1976, which announced the principle of autonomy of local authorities. The most important Portuguese legislative document defines a local authority as "(…) territorial collective persons endowed with representative bodies, which aim to pursue the interests of their respective populations." (Article 235,

Constitution of the Portuguese Republic). In the following article, CRP, it is stated that "on the mainland, local authorities are parishes, municipalities and administrative regions." Since the entry into force of article 238 of the Constitution of the Portuguese Republic, local authorities began to benefit from financial and patrimonial autonomy and tax powers. According to Portuguese law, municipalities have two representative bodies, the municipal assembly, which is the "(…) deliberative body of the municipality, which is made up of directly elected members in a greater number than the presidents of the parish council, which are part of it" and the municipal council, which is the "(…) collegiate executive body of the municipality."

In Portugal, Municipalities play a fundamental role in terms of social responsibility and are the public bodies most oriented towards their population, either because they are usually represented by local individuals or because they are characterized by being personalities with strong knowledge of local resources in its intervention zone. The study by Ribeiro et al. (2018) highlights the importance of the intervention of municipalities in the creation and promotion of strategies aimed at their development through social responsibility instruments. In this way, the dissemination of information by public bodies on this subject has a strong impact, as it allows these bodies to get closer to their citizens, thus showing their main stakeholders both their availability and their ability to serve them.

The disclosure of the commitments of public organizations to social responsibility can be carried out using various global references, but in this work, the analysis of the GRI reference and the King IV Report was privileged, which are characterized by including specific supplements whose guidance was specially directed to the Municipalities.

In addition to these organizations, it is important to highlight the role of Transparency International, whose main objectives include assessing the quality of governance, integrity, and transparency in central and local government. In Portugal, this organization prepared a Municipal Transparency Index (ITM). This Index is intended to measure "(…) annually the degree of commitment of the Local Authority to transparency (…)." That is, it allows citizens and decision-makers to verify the degree of transparency of Municipalities, through the analysis provided by them on their websites (Andrade & Batalha, 2017; Transparency International Portugal, 2020).

The European Union has been following up on the concerns about the subject under analysis, but only from 2013 onwards, did it start to recommend the disclosure of matters related to the "(…) environmental and social aspects of the company's activity necessary to understand the development, performance or the situation of the company." contained in the management report, according to § 26 of Directive 2013/34/EU. This guideline is consolidated with the publication of Directive 2014/95/EU, which had to be transposed into the legislation of the various Member States of the European Union by the end of 2016. As a result of this new guideline, large companies considered entities of public interest, or in large groups that, at the closing date of their individual or consolidated balance sheets, respectively, included an average of more than 500 employees, are now obliged to comply with the new disclosure requirements. It should be noted that, within the scope of application of the Directive, the public sector, in general, is not covered by the new law, only public companies are obliged to comply with it. This Directive was transposed into the Portuguese legal system through Decree-Law 89/2017 (DL) which requires that certain entities are required to include a non-financial statement in their accounts or prepare a separate report that includes the information required for that non-financial statement. The information requested in the non-financial statement must include environmental and social performance, measures aimed at workers, gender equality, non-discrimination, respect for human rights, and the fight against bribery and corruption.

The DL89/2017 also suggests that the elaboration of this information be done using international references such as the United Nations Global Compact, the guidelines of the Organization for Economic Cooperation and Development of multinational companies, the ISO 26000 Social Responsibility standard, or the GRI benchmark. This reference aims to guide companies to adopt the disclosure of environmental and social aspects in order to show their activity and progress. The main objective is to contribute to an analysis of the performance of companies and their impact on society, identifying their sustainability risks, reinforcing investor and consumer confidence, and allowing the comparison of information by large companies, from all sectors, in all Member States.

It should also be noted that although the designation adopted by the European Union for the disclosure of the aforementioned matters was the non-financial statement, its designation as sustainability report, social and environmental report, corporate social responsibility report, etc.

8.3 The Global Benchmarks of Social Responsibility

Since 1999, the Global Reporting Initiative (GRI) has published a set of guidelines for the preparation of sustainability reports. These guidelines present the principles and contents that must be implemented in the preparation of reports regardless of sector, size, or location. This set of guidelines aims to facilitate the understanding and management of the impacts of sustainability on the organization's activities and strategy through a vast number of performance indicators. The study carried out by KPMG (2017) states that the GRI guidelines continue to be the most used reference, voluntarily, throughout the world.

In order to understand the topic under discussion, it is also important to analyze the work carried out in South Africa by "The Institute of Directors in Southern Africa" (IoDSA) which established, in July 1993, the King Committee with a view to implementing corporate governance. more efficient. To prepare the South African economy for its participation in international markets, the King Committee produced its first code in 1994, which has been internationally recognized as one of the most comprehensive publications on the subject, presenting an inclusive approach to corporate governance. (IoDSA, 2020). After that code, three more King codes were published, the most recent version being the King Report IV.

In October 2016, GRI launched the first global standards for sustainability reporting called GRI Standards. These standards allow organizations to publicly report their economic, environmental, and social impacts and show how they can contribute to sustainable development, replacing the G4 version. Organizations that selected this benchmark had until July 1, 2018, for the transition from the G4 Guidelines to the new Standard, GRI Standards. The study developed is based on the GRI-G4 guidelines and not the current GRI Standards, for the simple fact that the Portuguese municipal councils, in 2018, still do not use them, and the differences between them are not very significant.

The GRI establishes that reporting principles are essential to ensure the transparency of the sustainability reporting process and, therefore, must be analyzed by all organizations. These principles are divided into two groups: (i) principles for defining report content and (ii) principles to ensure the quality of the report (GRI, 2015). It is necessary to consider the set of all principles, as it is essential for the effective transparency of the sustainability report process.

With regard to standard content, it can be divided into two subcategories: (i) general standard content, and (ii) specific standard content. This last subcategory is still divided into three groups, namely: (i) economic, (ii) environmental, and (iii) social, which are associated with the way in which the entities are managed. It is necessary to mention the existence of a hierarchy of elements for the dimensions of sustainability that consists of three elements: (i) categories: these are large areas or groupings of economic, environmental, or social aspects that affect stakeholders; (ii) aspects: these are elements related to a specific category, where a certain category may contain several aspects; and (iii) indicators: are the specific measures of a concrete aspect that can be used to monitor and demonstrate performance.

Performance indicators provide clarification on the organization's performance or economic, environmental, and social impacts, related to its material aspects (GRI, 2015). The G4 version, which represents the basis of the present study, contains 91 indicators that are grouped into three categories. For economic performance, there are nine indicators included in four aspects, for environmental performance, there are 34 indicators included in 12 aspects and for social performance, there are 48 indicators included in 30 aspects that, in turn, are grouped into several subcategories.

The King Code of Corporate Governance (King I), published in 1999, is one of the first codes with guidelines on corporate governance but did not mention sustainability (Eccles & Krzus, 2014). The King IV currently in force arises from the need to update the King III code, in November 2016, due to non-profit organizations, private companies, and public sector entities expressing difficulties in interpreting and adapting the King III to their particularities (Harduth & Sampson, 2016). Thus, at the time of its elaboration, King IV had this aspect present, choosing to cover organizations and government bodies, rather than just companies and boards of directors (IoDSA, 2016).

The King IV code resulted from the need to make information more accessible to all types of entities and sectors (Harduth & Sampson, 2016), as it contains fewer principles and these are more succinct, which facilitates their interpretation and implementation. This reference is also made up of sectoral supplements that help in its interpretation by organizations from different sectors.

The King IV code aims to: (i) promote corporate governance as an integral part of the administration of an organization and provide government results such as ethical culture, good performance, effective control, and legitimacy; (ii) broaden the acceptance of King IV, making it accessible and ready for implementation in a variety of sectors and organizational types; (iii) reinforce corporate governance as a holistic and interrelated set of agreements to be understood and implemented in an integrated manner; (iv) encourage the preparation of transparent and meaningful reports for stakeholders; and (v) present corporate governance concerned not only with structure and process but also with ethical and conscientious conduct. It should be noted that the governance body is the main point of corporate governance, thus making it the main theme of King IV. Its promotion is made in order to be applied to all organizations, due to one of its objectives being the application and acceptance of corporate governance, this application becomes accessible and suitable for all sectors and organizations (IoDSA, 2016).

Among one of the six components of the King Report, there are also sectoral supplements that were included in King IV in order to aim at expanding and accepting corporate governance, making it accessible and suitable for different sectors, organizations, and entities of different sizes, resources, and complexity, in addition to the traditional public of companies listed on the stock exchange, public and large private companies. The industry supplement determines how concepts and recommendations can be interpreted, adapted, and customized to meet the needs and requirements of organizations in specific industries. Its composition presents a sectoral supplement for municipalities, non-profit organizations, pension funds, small and medium-sized companies, and state entities (IoDSA, 2016). The present study will only analyze the sectoral supplement for municipalities.

The social responsibility report is one of the most used ways by organizations to communicate their practices and results, due to the growing concern with sustainable development and its environmental, economic, and social aspects. The two references presented above are considered the most relevant from a global perspective. For the preparation of these reports, in the public sector, it is essential to use sectorial supplements, they operate as a complementary tool in the analysis of risks and opportunities (Barsano & Barbosa, 2014, 2019), giving a more adequate response to the realization of these reports in the public sector.

GRI recognized a need to create industry-specific guidelines to complement the general information raised in the guidelines. This becomes essential in order to allow robust and useful reports, and also to increase the applicability and acceptance of GRI guidelines across sectors and around the world (GRI, 2005).

The main objectives of sustainability reporting are to help public bodies gain a broader view of their own sustainability activities. Often, the public body's main function will be related to an economic, environmental, and social element and its performance reports will reflect the same. GRI-compliant reporting provides a framework to help public bodies report these impacts, as well as the interactions between different aspects of their activities in an integrated manner (GRI, 2005).

Public bodies must report their sustainability performance in all aspects of the "triple bottom line," providing both qualitative and quantitative information, which can be divided into three types (GRI, 2005) of information: organizational performance; public policies, and implementation measures; and about the context or state of the environment. The first type of information concerns what can be reported through the use of performance indicators. This type of information illustrates the organization's internal policies and its role as a consumer and employer. The second type lists information on public policies and implementation measures, these are related to sustainable development and its performance. The last type addresses information about the context or state of the environment, focusing on economic, environmental, or social issues within the conditions of the mandate or area of jurisdiction of the sector that may have in view public policies and implementation measures (GRI, 2005).

Since the publication of the public sector supplement, GRI has initiated a study to analyze its use and the level of disclosure of its items. This assessment concluded that organizations use more of the general structure of the GRI guidelines to the detriment of the supplement (GRI, 2010). According to the organization, in preparing the sustainability report, municipalities must follow specific guidelines for the public sector. As the use of the supplement is complementary to the information required by the general guidelines, a structured report is obtained, but with specific lines relating to the public sector.

For the application of the King IV code, there is a specific supplement for application to municipalities. This, instead of repeating the King IV code in full, explains how the terminology used in King IV can be exchanged for the appropriate terms for municipalities, where it is emphasized that the leadership structure of municipalities is not comparable to other organizations (IoDSA, 2016).

Municipalities are autonomous but operate within a cooperative governance system. This cooperative governance requires that the various spheres of government (national, provincial, and local) cooperate in the exercise of their powers and in the performance of their functions. Municipal governance must be seen in the context of cooperative governance without harming municipal autonomy (IoDSA, 2016).

It should be noted that the GRI supplement is not for sectoral application, but to complement the guidelines. This is not the case in the supplement for the municipal sector of King Report IV, in which the Municipalities can restrict themselves to the application of the same.

Although the GRI supplement is constituted globally for the public sector, it expresses the information that needs to be reported by these bodies, namely: (i) organizational performance; (ii) information on public policies and implementation measures, and (iii) information about the context or state of the environment. The last piece of information, information about the context or state of the environment, is usually the subject of the highest level of disclosure by the municipalities, as mentioned earlier in the study carried out by the GRI in 2010. This supplement consists of (i) comments on the disclosure and performance elements existing in the GRI indicators; (ii) a new set of disclosure elements on public policies and implementation measures; and (iii) new performance indicators in the economic and social sections. Thus, the structure of the GRI supplement is composed of: Vision and Strategy; Organizational Profile; Governance Structure and management systems; Performance indicators; and Public policies and implementation measures.

The study carried out by KPMG (2017), previously mentioned, states that the GRI guidelines continue to be the most used reference voluntarily around the world. But it is up to each municipality to choose which benchmark it wants to be applied and, in turn, the supplement, as both are voluntary in nature.

The application of the social responsibility report, although not legally required, is a great benefit, as it discloses to the stakeholders the actions carried out, such as the reduction of environmental impacts, the programs related to the well-being of the community and the practices adopted, as well as helping municipalities to identify their weaknesses and their options for achieving a higher level of social responsibility.

8.4 The Research Design

The main objective of this study is to analyze and understand what the municipalities disclose in terms of (i) General information; (ii) Social information; (iii) Information on contracting services and public works; (iv) Economic information, and (v) Environmental information, and what factors may drive this disclosure. Studies on social responsibility in the public sector, such as those prepared by Nevado-Gil and Gallardo-Vázquez (2016) and Navarro et al. (2010), suggest that the existence of several factors such as size, dependent population, unemployed population, education level, institutional capacity, level of indebtedness, political competence, and electoral participation affect the level of disclosure on social responsibility. Thus, based on these studies, we analyzed the influence of these factors, specifically considering the sociodemographic, socioeconomic, fiscal, and political dimensions.

The sociodemographic dimension analyzes the relationship between the size of the municipality and the dissemination of information. Some studies analyze the size of the municipality using the analysis of the population size factor (Ahmed & Courtis, 1999; Bastida & Benito, 2007; Frost & Seamer, 2002; García-Sánchez et al., 2013; Navarro et al. 2008; Navarro Galera, et al., 2011; Navarro et al., 2010; Nevado-Gil & Gallardo-Vázquez, 2016; Serrano-Cinca et al., 2009a, 2009b). Studies by Ahmed and Courtis (1999), García-Sánchez et al. (2013), Navarro Galera et al. (2011), Nevado-Gil and Gallardo-Vázquez (2016), and Serrano-Cinca et al. (2009a) demonstrate a significant and positive relationship between the dissemination of information about social responsibility and its dimension. That is, municipalities that are demonstrably more populous are more likely to disseminate information than smaller ones.

In addition to the total size of the municipality, the structure of this population can also condition disclosure requirements. Studies made by Navarro et al. (2010), Navarro Galera et al. (2011), and García-Sánchez et al. (2013) argue that larger municipalities, by encompassing a higher number of people and diversity, in addition to being usually made up of more qualified individuals who are concerned about the demand for information about social responsibility, can encourage greater practices of disclosure of social responsibility.

The age structure of the population is another factor considered to influence the level of disclosure of social responsibility. The population younger than 19 years and older than 65 years is considered the most dependent population, and when this proportion of people is greater in the total weight of the population, this characteristic means a significant increase in public expenditure. The investigation by Jorge et al. (2011) shows evidence supporting that the larger the population at these ages means greater the demand of citizens regarding social responsibility.

Some studies link the socioeconomic dimension with transparency. Research by Piotrowski and Van Ryzin (2007) and Guillamón et al. (2011a, 2011b) show that this dimension is crucial for increasing transparency.

The unemployment rate also represents the economic situation of the municipality as it results from the economic performance of the municipality, so its relationship with transparency is also a factor to analyze (Caamaño-Alegre et al., 2013; Cuadrado-Ballesteros, 2014; Guillamón et al., 2011b; Navarro-Galera et al., 2011; Río & Toharia, 2009; Sol, 2013). Several studies on transparency in Spanish municipalities show that municipalities with a higher unemployment rate are associated with municipalities with a lower level of transparency and lower disclosure rates (Guillamón et al., (2011a, 2011b); Sol (2013) and Caamaño-Alegre et al. (2013)).

The investigations by Cruz et al. (2012), Navarro-Galera et al. (2011), and Piotrowski and Van Ryzin (2007) relate to transparency and the level of disclosure to education and conclude that there is evidence that education and transparency are positively related. Navarro-Galera et al. (2011) also argue that stakeholders with a higher level of education feel more comfortable using the internet and have more confidence to request information, concluding that a higher level of education favors greater publication of economic information.

Any innovation or reform in the public sector is closely linked to the level of government resources available. This level of resources corresponds to the institutional capacity to obtain resources, fiscal dimension, and measures the municipal capacity to provide services to its citizens (García-Sánchez et al., 2013). The study of Navarro et al. (2010) considers that municipalities with a larger budget also guarantee more resources to improve the usefulness of their information systems. This could even give rise to additional incentives for greater dissemination of information on social responsibility.

The studies of García-Sánchez et al. (2013) and Guillamón et al. (2011a, 2011b) found that municipalities that receive more taxes and transfers disclose more financial information, and are therefore classified as more transparent. In the same research line, Guillamón et al. (2011b) state that, given this transparency, the most financially powerful municipalities are those that also have higher per capita expenditures. However, Navarro et al. (2010) and García-Sánchez et al. (2013) do not find significant evidence that disclosure rates are favored by the greater volume of budgetary expenditures.

Some studies also analyze the relationship between the level of municipal debt and transparency (Alt & Lassen, 2006; Alt et al., 2006; Sol, 2013) and find that more fiscal transparency is associated with lower debt levels. However, Sol (2013) does not find a significant relationship between the level of indebtedness and transparency, but the US-based study by Smith (2004) shows different evidence in finding a significant relationship between the level of indebtedness and transparency.

García and García (2008) and Prado, García, and Cuadrado (2010) identify that, in addition to the previous dimensions, a political dimension can be identified, which relates political competence to the level of information dissemination. In their studies, these researchers identify a positive and significant relationship between the level of political competition and the degree of disclosure of social responsibility.

The studies justify this relationship because as political competition increases, so does the level of pressure exerted and, with that, the party in power seeks to demonstrate to citizens its ability to offer greater benefits than competitors. To demonstrate this competence, it is necessary to disclose and act with transparency (García & García, 2008). Consequently, the level of citizen participation in elections demonstrates their interest (Esteller-Moré & Otero, 2012; Hollyer et al., 2011), constituting an analysis variable for information disclosure and transparency (Caamaño- Alegre et al., 2013; Sol 2013).

Guillamón et al. (2011b) suggest that the level of transparency is higher when citizens have a greater political commitment to participating in the electoral act. However, the study by Sol (2013) does not find significant statistical evidence in this regard. It should be noted that citizens can also increase voter turnout if the level of transparency and disclosure is greater, as it can convey confidence to them, and in this way can change perceptions about politicians (Fung, 2013).

After identifying the different dimensions that condition the disclosure of information on social responsibility by public entities, and based on the studies of Nevado-Gil and Gallardo-Vázquez (2016) and Navarro et al. (2010), we formulate the following research hypotheses.

H_1: Population size encourages SR information dissemination practices.
H_2: The dependent population encourages SR information dissemination practices.
H_3: The unemployed population does not encourage SR information dissemination practices.
H_4: The municipality's level of education encourages the practices of disseminating information about RS.
H_5: Institutional capacity encourages SR information dissemination practices.
H_6: The level of indebtedness does not encourage SR information disclosure practices.
H_7: Political competence encourages SR information dissemination practices.
H_8: The level of electoral participation encourages SR information dissemination practices.

To carry out the study, based on the literature review, the variables presented in Table 8.1, which also describes the methodology used for its measurement, were used.

Table 8.1 Variables and their measurement

Variables	
Population size (H_1)	Number of inhabitants per municipality
Dependent population (H_2)	Number of inhabitants aged < 19 and > = 65 per municipality
Unemployed population (H_3)	Number of unemployed inhabitants by municipality
Education level (H_4)	Number of inhabitants with higher education by municipality
Institutional capacity (H_5)	Expenditure on the staff of municipal councils per capita
Indebtedness level (H_6)	Municipal debt expressed in euros per inhabitants of the municipality
Political competence (H_7)	Number of political parties participating in the 2017 municipal elections
Electoral participation (H_8)	Abstention in the last municipal elections

The present investigation aims to analyze the information reported by Portuguese municipalities on social responsibility and to identify the main factors that influence these dissemination practices. In the first stage of the study, the selection of municipalities included in the sample was carried out, having been selected by all Portuguese municipalities related to the North region of Portugal. To delimit the study sample, the NUTS (Nomenclatures of Territorial Units for Statistical Purposes) was used as a criterion, which defines the statistical sub-regions of the territory of the European Union countries. In Portugal, these regions are subdivided into three levels: NUTS I, NUTS II, and NUTS III (CCDR, 2020). Applying the NUTS to the study sample, the Northern region of Portugal is positioned at the second level (NUTS II), with 86 Municipalities being included here. Thus, the study sample represents 31% of all municipalities in mainland Portugal, where almost 35% of the population residing in Portugal is concentrated, which accounts for close to 39% of Portuguese exports and represents about 29% of Portuguese GDP.

The preparation of the proposed study began with the collection of information on social responsibility, using the analysis of web pages relating to all selected municipalities. The content analysis was carried out based on the observation of the accounts reports, the management reports and the sustainability report, and other documents that could complement the required information, according to the methodology used by Navarro et al. (2010) and Nevado-Gil and Gallardo-Vázquez (2016). This analysis was also used in research work on the dissemination of social responsibility carried out, among others, by Guthrie and Farneti (2008), Joseph (2010), Joseph and Taplin (2011), Navarro et al. (2010), Navarro Galera et al. (2011), Ribeiro and Guzmán (2011), and Ribeiro et al. (2018).

In order to analyze the dissemination practices of the municipalities in the North of Portugal, during the year 2020, 100 dissemination items were analyzed on the 86 websites of the municipalities in the sample. These 100 indicators were divided into five categories of information, based on the framework of indicators developed by Nevado-Gil and Gallardo-Vázquez (2016) and Navarro et al. (2010) and GRI Guidelines (GRI, 2015), as the particularities of the public sector and incorporating the

Table 8.2 Disclosure indicators by type of information

1	General information (25 items)
2	Social information (19 items)
3	Information on contracting services and public works (17 items);
4	Economic information (20 items)
5	Environmental information (19 items)

items contained in the Municipal Transparency Index published in 2017 (Andrade & Batalha, 2017). Table 8.2. shows the breakdown of indicators.

Regarding the criterion used to classify the presence or absence of information regarding each of the indicators, a dichotomous scale was used: 1, in case the indicator was disclosed, and 0, in case it was not disclosed. This scoring method, despite not considering the quality and quantity of information disclosed, has been used in several studies such as Ettredge et al. (2001), Bastida and Benito (2007), García-Sánchez et al. (2013), and Nevado-Gil and Gallardo-Vázquez (2016).

Subsequently, in order to measure the level of disclosure of information on social responsibility of municipalities, a disclosure sub-index was created for each category of information that covers the information items of each category:

$$IDX_j = \frac{\sum_{i=1}^{B}(Aij)}{M}[1]$$

where:

IDX$_j$: Information Dissemination Index by dimension/category X, in the municipality "j."
Ai (j) = 1: If the indicator characteristic (i) is present in the municipality (j).
Ai (j) = 0: If the indicator characteristic (i) is not present in the municipality (j).
B: Total score obtained by each municipality in each of the information categories.
M: Number of items/indicators that make up each dimension/category X.
X: Each of the dimensions/categories that make up the Disclosure Index.
IDG: General Dimension of the Municipality; SDI: Social Dimension; IDC: Dimension of Contracting Services and Public Works; SDI: Economic Dimension; IDA: Environmental Dimension.

The option of using indexes to measure the level of dissemination of information on this topic has been widely used in several studies such as Joseph (2010), Joseph and Taplin (2011), Navarro et al. (2010), Navarro Galera et al. (2011), Nevado-Gil and Gallardo-Vázquez (2016), and Sciulli (2009). Following the same methodology, a total disclosure index on social responsibility (IDRS) was also created, whose function is to aggregate information regarding the various sub-indices of the five dimensions. Considering that the aim here is to interpret which aspects are or are not disclosed by the municipalities, an equal weight was assigned to the five dimensions (20%), in order to determine whether there is a balance according to the proposed dimensions. In order to reconcile this information, the IDRS calculation formula is presented below:

$$IDRS_j = (IDG_j * 0,2) + (IDS_j * 0,2) + (IDC_j * 0,2) + (IDE_j * 0,2) + (IDA_j * 0,2)$$

where: $IDRS_j$ is the Total Disclosure Index on Social Responsibility; IDG is the General Dimension of the Municipality; IDS is the Social Dimension, IDC is the Dimension of Contracting Services and Public Works; IDE is the Economic Dimension, and IDA is the Environmental Dimension.

The choice of the analyzed municipalities was based on the importance of the Northern Region of Portugal for the Portuguese economy and, as far as we are aware, the lack of studies on this subject concerning this Portuguese region.

In order to analyze the research hypotheses, we used different statistical methodologies. We used the multivariate technique of Cluster Analysis as several studies such as Mar-Molinero and Serrano-Cinca (2001), Navarro et al. (2010), and Nevado-Gil and Gallardo-Vázquez (2016), considering as variables for the segmentation of municipalities the five disclosure sub-indices.

Finally, to assess the possible association between the proposed factors and the disclosure rates of each category, we used the inferential statistical procedure of comparing the averages to the quantitative variables in the different clusters established by the categorical variable resulting from cluster analysis. This comparison of means is performed using Analysis of Variance (ANOVA). If any of the ANOVA application assumptions are not met, alternatively, the non-parametric Kruskal–Wallis test can be used.

8.5 Analysis and Discussion of Results

The type of information disclosed by the 86 municipalities was analyzed considering the following categories: (i) General information; (ii) Social information; (iii) Information on contracting services and public works; (iv) Economic information; and (v) Environmental information.

In an analysis of the global levels of disclosure, in percentage, we found that the set of municipalities analyzed discloses only 50.88% of the information about social responsibility in relation to all the items analyzed. For an analysis of the global results, by categories, see Fig. 8.1, which allows us to conclude that the category of economic information is the one with the highest level of disclosure, followed by the category of social information, general information, and information on contracting

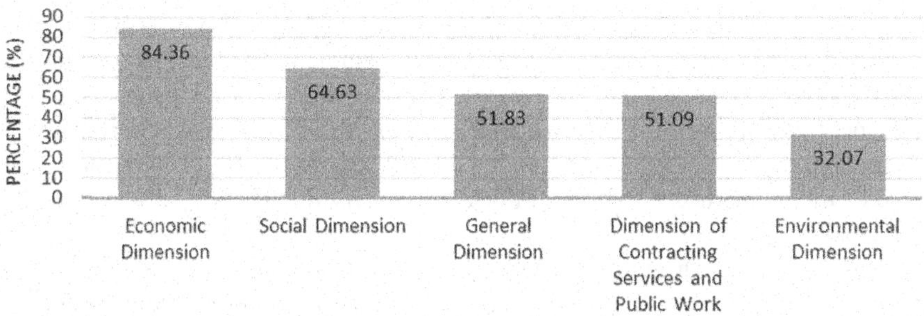

Fig. 8.1 Social responsibility disclosure percentage by information categories

services and public works. Finally, the one with the lowest level of disclosure is the category related to environmental information.

The global index obtained a total value of 0.43, which comes from the sum of five categories multiplied by 20% for each one (0.10 + 0.01 + 0.09 + 0.16 + 0.06). In an analysis of the disclosure index by a municipality, Bragança is the municipality with the highest total disclosure rate (0.74), followed by Porto (0.67) and Vila Nova de Cerveira (0.64). At the opposite extreme are Penedono (0.21), Vinhais (0.21), and Sernancelhe (0.26) with the lowest disclosure rate.

In the general category, the municipality with the highest disclosure is Bragança (0.19). The results obtained also show that only 22 Municipalities (25.58%) identify the importance of sustainability and only 16 (18.60%) declare which failures characterize them. In the part of the municipal structure, there is a good level of disclosure, with 80 municipalities (93.02%) disclosing this information. However, in the dissemination of information about the different municipal bodies and their functions, only 54 Municipalities (62.79%) disclosed that information.

In the social category, the possibility of the municipalities' websites having a mechanism for hearing the entities was analyzed, but only five municipalities (5.81%) have this concern. Most municipalities (98.84%) allow internal research on the website page and have links to social networks (95.35%), with the existence of a mailbox for people to make complaints and suggestions (93.02%). Note that all municipalities highlight news on their website. The possibility of performing administrative procedures or requesting licenses online is only present in 65.12% of the municipalities.

In the category of contracting services and public works, in terms of public procurement, it appears that approximately half of the municipalities disclose public procurement procedures and the respective documentation involved. Still in relation to this item, only 5.81% (5) municipalities publish audit reports and opinions from inspection bodies. In terms of spatial planning, there is good general dissemination, where 94.19% (81) include a section on this topic.

In the economic category, it appears that the vast majority of entities disclose accounting documents such as the budget, balance sheet, income statement, management report, cash flow and budget statements, and the annual execution of the

Multiannual Investment Plan. Regarding municipal debt, this also presents a significant level of disclosure, with 96.51% (83) of the municipalities making the disclosure. It should be noted that the list of debts to suppliers, with the respective delay periods, is only published by 59.30% (51) of the municipalities. Regarding income and expenses, 98.84% (85) of the entities disclose them. With regard to the value of municipal taxes and fees, information is disclosed by 89.53% (77) of the municipalities.

Finally, in the environmental category, there is a low level of disclosure, with only two municipalities (2.33%) presenting updated information at this level, and only seven (8.14%) disclosing information on the total expenditures and investments under this heading. Regarding the reduction in energy consumption and air and noise pollution, little disclosure remains, with six (6.98%) and four (4.65%) municipalities, respectively, disclosing information.

However, the dissemination of actions to improve energy consumption (80.23%), mitigate environmental impacts (89.53%), and promote citizens' sensitivity (93.02%) is significantly higher.

Regarding environmental policies, only 10.47% (9) municipalities refer to them. Disclosure of sanctions and non-compliance with environmental legislation is 5.81% (5). With regard to garbage collection, 43.02% (37), and recycling 36.05% (31), of the municipalities disclose this information.

The second analysis aspect of this investigation is the identification of the factors that influence the disclosure rates obtained in the sample under study. To achieve this objective, we chose to apply cluster analysis, using a non-hierarchical technique that resulted in 3 groups, that is, the 3 clusters represented in Table 8.3. This procedure has been used in several studies, for example, by Mar-Molinero and Serrano-Cinca (2001), Navarro et al. (2010), and Nevado-Gil and Gallardo-Vázquez (2016).

Table 8.3 Municipalities grouped by cluster

Cluster 1	Alfândega da Fé; Arcos de Valdevez; Armamar; Bragança; Chaves; Guimarães; Lamego; Maia; Marco de Canaveses; Matosinhos; Moimenta da Beira; Oliveira de Azeméis; Paredes; Penafiel; Ponte do Lima; Porto; Póvoa de Varzim; Santa Maria da Feira; Santo Tirso; São João da Pesqueira; São João Madeira; Valongo; Valpaços; Vila Nova de Cerveira
Cluster 2	Amarante; Amares; Arouca; Baião; Boticas; Braga; Caminha; Castelo de Paiva; Cinfães; Espinho; Gondomar; Lousada; Macedo de Cavaleiros; Melgaço; Mesão Frio; Mondim de basto; Montalegre; Murça; Paços de Ferreira; Póvoa de Lanhoso; Resende; Sabrosa; Tarouca; Torre de Moncorvo; Trofa; Valença; Viana do Castelo; Vila do Conde; Vila Nova de Famalicão; Vila Nova de Gaia; Vila Pouca de Aguiar; Vila Real
Cluster 3	Alijó; Barcelos; Cabeceiras de Bastos; Carrazeda de Ansiães; Celorico de Basto; Esposende; Fafe; Felgueiras; Freixo de Espada à Cintra; Miranda do Douro; Mirandela; Mogadouro; Monção; Paredes de Coura; Penedono; Peso da Régua; Ponte da Barca; Ribeira de Pena; Santa Marta de Penaguião; Sernancelhe; Tabuaço; Terras de Bouro; Vale de Cambra; Vieira do Minho; Vila Flor; Vila Nova de Foz Côa; Vila Verde; Vimioso; Vinhais

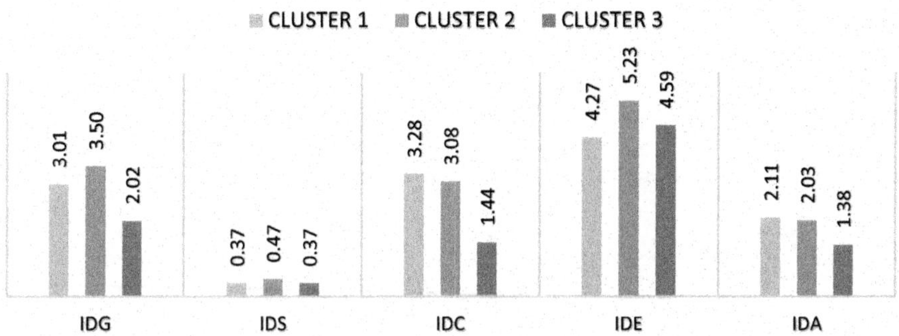

Fig. 8.2 Average of the indices, by cluster and by category

Considering the clusters described in Fig. 8.2 shows the average of the various disclosure sub-indices for the three clusters.

From this graph, it can be concluded that cluster 3 has a lower overall rate of contracting services and public and environmental works; cluster 1 has a lower social and economic index; in turn, cluster 2 has a higher general, social, and economic index, while cluster 1 has a higher rate of contracting services and public and environmental works. Still regarding the average of the indices, per cluster, in particular, the total index, it can be seen through Table 8.4 that cluster 1 has the highest average, followed by cluster 2 and, finally, cluster 3.

Then we use the normality test, more specifically, the Kolmogorov–Smirnov test (Table 8.5).

The test represented in Table 8.6 was interpreted considering the significance level of 5%. The normality of the data is one of the assumptions for the application of ANOVA; however, given the test results (Sig = 0), the normality of the variables cannot be assumed.

Table 8.6 show the Kruskal–Wallis tests, considering the significance level of 5% and using asymptotic significance.

As the hypothesis of normality of the variables is not verified, the non-parametric Kruskal–Wallis test was applied. The data obtained suggest that the behavior of the variable's population size, dependent population, unemployed population, level of education, political competence, institutional capacity, and level of electoral participation differ between the various clusters. As for indebtedness, no significant differences were found for the 3 clusters, so we should accept the null hypothesis. In this

Table 8.4 Average of the total index per cluster

IDRS	Mean	Min	Max
Cluster 1	0.543	0.475	0.745
Cluster 2	0.438	0.384	0.519
Cluster 3	0.338	0.205	0.433

Table 8.5 Kolmogorov–Smirnov tests

Kolmogorov–Smirnov Tests	Population size (H_1)	Dependent population (H_2)	Unemployed population (H_3)	Education level (H_4)	Institutional capacity (H_5)	Indebtedness level (H_6)	Political competence (H_7)	Electoral participation(H_8)
Statistical test (Lilliefors significance	0.246	0.243	0.278	0.305	0.137	0.188	0.229	0.249
Sig	0.000	0.000	0.000	0.000	0.000	0.000	0.000	0.000

Table 8.6 Kruskal–Wallis tests

Independent samples Kruskal–Wallis tests	Population size (H_1)	Dependent population (H_2)	Unemployed population (H_3)	Education level (H_4)	Institutional capacity (H_5)	Indebtedness level (H_6)	Political competence (H_7)	Electoral participation (H_8)
Sig	0.001	0.001	0.001	0.002	0.018	**0.566**	0.001	0.001

way, based on the differences identified, for each variable, the clusters are identified (Table 8.7). Subsequent, Table 8.8, some descriptive statistics are presented for the factors under study: size, dependent population, unemployed population, level of education, institutional capacity, indebtedness, political competence, and level of electoral participation; for the total index; and for various clusters.

When analyzing the population size factor, we did not find differences between clusters 1 and 2, but there are differences when comparing with cluster 3. The values presented, mean rank, indicate that the population size is smaller in the 3rd cluster. Thus, with the results obtained in clusters 1 and 2 showing a higher average population size and also a higher total disclosure rate, it seems that it is possible to validate hypothesis: H1: "The size of the entity favors information disclosure practices about RS." Thus, larger municipalities assume greater responsibility for disseminating social responsibility information. This result is in line with those obtained by other authors, such as Nevado-Gil and Gallardo-Vázquez (2016) and Navarro et al. (2010).

When the variable analyzed is the dependent population, we did not observe differences between clusters 1 and 2, but these are verified if we make the comparison with cluster 3. By analyzing the mean, it is verified that the dependent population is smaller in the third cluster. The municipalities grouped in clusters 1 and 2, contain a larger dependent population and assume a greater responsibility for disseminating social responsibility information in relation to smaller ones, so we are led to validate hypothesis H2: The number of dependent populations favors the practices of dissemination of information about social responsibility. This result corroborates the conclusions of Nevado-Gil and Gallardo-Vázquez (2016) and Navarro-Galera et al. (2011).

When we analyze the unemployed population variable, we verify that there are no differences between the unemployed population for clusters 1 and 2. The mean rank indicates that the unemployed population is smaller in the 3rd. cluster. The statistical significance obtained does not allow for validating the research hypothesis H3, so the municipalities with the largest unemployed population assume greater responsibility for disseminating social responsibility information in relation to the smaller ones. Authors such as Caamaño-Alegre et al. (2013); Guillamón et al. 2011a, 2011b) and Sol (2013) obtained the same results.

For the variable level of education, the results obtained lead to the validation of hypothesis H4, which is supported by the results showing that municipalities with a higher level of education assume greater responsibility in disseminating information on social responsibility than smaller ones, which is in line with the study by Nevado-Gil and Gallardo-Vázquez (2016).

Based on the analysis of the results, we cannot conclude that the greater the institutional capacity, the lower the disclosure rate, and hypothesis H5 cannot be validated. The results obtained in the study developed do not corroborate those obtained in the reference study.

By applying the Kruskal–Wallis method, there is evidence that the behavior of the indebtedness variable is identical for the three clusters. In the absence of significant differences between the three clusters, it does not seem reasonable to assume the

Table 8.7 Homogeneous Subsets (Kruskal–Wallis method)

	Pop Size		Dependent population		Unemployed population		Education level		Institutional capacity		Political competence		Electoral participation level		
Subset	S 1	S 2	S 1	S 2	S 1	S 2	S 1	S 2	S 1	S 2	S 1	S 2	S1	S2	S3
Cluster 3	30,759		31,103		30,466		31,655		33,083		34,207		30,379		
Cluster 2		45,061		44,788		46,606		44,576		43,030	41,242			45,303	
Cluster 1		56,750		56,708		54,979		56,333		52,655		57,833			56,875
test statistics		3,455		3,515		1,887		3,336		2,258	1,442				
Sig (bilateral test)		**0,063**		0,**061**		0,169		**0,068**		0,133	0,230				

Table 8.8 Mean rank and IDRS mean

Cluster	Population Size		Dependent population		Unemployed population		Education level		Institutional capacity		Indebtedness level		Political competence		Electoral participation level	
	Mean rank	IDRS mean	Mean rank	IDRS mean	Mean rank	IDRS mean	Mean rank	IDRS mean	Mean rank	IDRS mean	Mean rank	IDRS mean	Mean rank	IDRS mean	Mean rank	IDRS mean
1	64,706,29	0,543	25,344,00	0,543	2405,17	0,543	7745,92	0,543	251,11	0,543	613,29	0,543	5,17	0,543	24,834,29	0,543
2	44,886,91	0,438	17,148,06	0,438	1814,67	0,438	4637,88	0,438	300,19	0,438	620,48	0,438	4,27	0,438	17,042,33	0,438
3	18,564,28	0,338	7320,86	0,338	601,00	0,338	2128,97	0,338	366,49	0,338	677,41	0,338	3,90	0,338	6594,66	0,338

Table 8.9 Summary of research hypotheses results

H1: Population size encourages SR information dissemination practices	**Validated**
H2: The dependent population encourages SR information dissemination practices	**Validated**
H3: The unemployed population does not encourage SR information dissemination practices	Not Validated
H4: The municipality's level of education encourages the practices of disseminating information about RS	**Validated**
H5: Institutional capacity encourages SR information dissemination practices	Not Validated
H6: The level of indebtedness does not encourage SR information disclosure practices	Not Validated
H7: Political competence encourages SR information dissemination practices	**Validated**
H8: The level of electoral participation encourages SR information dissemination practices	**Validated**

validity of hypothesis H6. The results obtained are in accordance with those of the authors Nevado-Gil and Gallardo-Vázquez (2016).

The hypothesis "H7: Political competence favors the practices of disseminating information on SR," is validated based on the results obtained, since municipalities with greater political competence assume greater responsibility in disseminating SR information in relation to smaller ones. The results obtained do not agree with those of the study by Nevado-Gil and Gallardo-Vázquez (2016).

Hypothesis H8, in which the level of electoral participation favors information dissemination practices on social responsibility, is validated, as the greater the electoral participation, the greater the dissemination of information. An identical result is obtained by the authors Nevado-Gil and Gallardo-Vázquez (2016).

After all the analyses are performed, we obtained the following results, shown in Table 8.9.

The results obtained allowed us to validate five of the research hypotheses initially defined, allowing us to conclude that greater practices of dissemination of social responsibility in the municipalities of the Northern Region of Portugal are associated with the size of the municipality, the importance of its dependent population, the level of education of the population of that municipality, political competence, and the level of electoral participation. These results are in accordance with the stakeholder theory, in which the dissemination of information on social responsibility is an instrument used by municipalities to show that they are acting in accordance with the expectations of these same stakeholders (citizens). A similar consideration can be made about the theory of legitimacy with this disclosure of information being a tool to legitimize and confirm the choice of municipal governance.

8.6 Conclusions

This investigation analyzes and quantifies the matter of disclosure of social responsibility in 86 Municipalities belonging to the North region of Portugal, using the analysis of their web pages and investigating the possible explanatory factors for their level of disclosure. After identifying the explanatory factors, in which the sociodemographic, socioeconomic, fiscal, and political factors were analyzed, a social responsibility disclosure index (IDRS) was elaborated.

The results obtained suggest that the disclosure level of responsibility is medium, as can be seen with a result of 56.77% of the social responsibility items to be disclosed by the municipalities. The category with the greatest disclosure is the economic one, followed by the social category, the general category, public procurement, and, with the least disclosure, the environmental category was identified. This last category has a visible lack of disclosure on web pages, so it was identified that these institutions should be particularly concerned with the dissemination of information on their performance in environmental aspects that are considered increasingly crucial.

Regarding the factors associated with dissemination, these suggest that the dissemination of information on social responsibility is related to the size of the population, the proportion of its dependent population, the level of education, political competence, and, also, the level of participation in the elections. However, the same association was not visible with the number of unemployed population, with the institutional capacity, or with the level of indebtedness.

It should be noted that the composition of the sample is identified as the main limitation of the present investigation, as it includes only the municipalities belonging to the Northern Region of Portugal. In the future, it would be of great interest to apply the same study to the remaining Portuguese municipalities, hoping to identify which Portuguese region has the highest IDRS.

This study mainly aims to contribute to reducing the gap in research in the area of disclosure of social responsibility information in the case of the Portuguese municipalities. Among its main contributions, we highlight the identification of the need to improve the performance of municipalities in terms of disseminating social responsibility, as well as encouraging those responsible to improve the dissemination of information on their web pages, mainly at the level of disclosures on environmental matters.

References

Ahmad, Z., Sallet H., & Mohammad, J. (2003). Determinants of environmental reporting in Malaysia. *International Journal of Business Studies*, 69–90.

Ahmed, K., & Courtis, J. K. (1999). Associations between corporate characteristics and disclosure levels in annual reports: A meta-analysis. *British Accounting Review, 31*, 35–61.

Alt, J. E., Lassen, D. D., & Rose, S. (2006). The causes of fiscal transparency: Evidence from the U.S. *States. IMF Staff Papers, 53*, 30–57. Disponível em https://link.springer.com/article/10.2307/30036021, consultado em 14 de fevereiro de 2020.

Alt, J. E., & Lassen, D. D. (2006). Fiscal transparency, political parties, and debt in OECD countries. *European Economic Review, 50*(6), 1403–1439. ISSN: 0014–2921.

Andrade, C. & Batalha, J.P. (2017). *ITM-Índice de Transparência Municipal -Apresentação e Indicadores, Transparência e Integridade, Associação Cívica.* Disponível em https://transparencia.pt/wpcontent/uploads/2017/11/ITM_Apresentacao_e_Indicadores_2017.pdf , consultado em 5 de janeiro de 2020.

Bandeira, A. M. A. (2010). Os Desafios e as Dificuldades da Contabilidade Face à Responsabilidade Social das Empresas. *XIV Encontro AECA*, Coimbra. http://www.aeca1.org/pub/on_line/comunicaciones_xivencuentroaeca/cd/6h.pdf, 22 out. 2019.

Barsano, P. R., & Barbosa, R. P. (2014). *Gestão ambiental.* São Paulo. ISBN:9788536528830.

Barsano, P. R,, & Barbosa, R. P. (2019). *Meio Ambiente: Guia Prático e Didático.* (3ª ed.). São Paulo. ISBN:978–85–365–3224–0.

Bastida, F., & Benito, B. (2007). Central government budget practices and transparency: An international comparison. *Public Administration, 85*(3), 667–716.

Bilhim, J. (2004). *A Governação nas autarquias locais*. Porto: Sociedade Portuguesa de Inovação (SPI). ISBN: 972–8589–37–9.

Bowen, H. R. (2013). *Social responsibilities of the businessman.* University of Iowa Press. ISBN: 978_I-60938-206–4.

Burrit, R. L., & Welch, S. (1997). Australian commonwealth entities: An analysis of their environmental disclosures. *Abacus, 33*(1), 69–87. https://doi.org/10.1111/1467-6281.00004

Caamaño-Alegre, J., Lago-Peñas, S., Reyes-Santias, F., & Santiago-Boubeta, A. (2013). Budget transparency in local governments: An empirical analysis. *Local Government Studies, 39*(2), 182–207. https://doi.org/10.1080/03003930.2012.693075

Carroll, A. B. (1999). Corporate social responsibility: Evolution of a definitional construct. *Business & Society., 38*(3), 268–295. https://doi.org/10.1177/000765039903800303

Caulkins, D., Evans, B., Liang, C., & Chaet, J. (2013). Howard bowen & corporate social responsibility. *The Grinnell Magazine, Winter, V, 46*(2), 20–21.

CCDR Norte. (2020). *Regulamento (UE) nº868/2014 de Comissão.* Disponível em https://www.ccdr-n.pt/pagina/regiao-norte/apresentacao, consultado em 5 de janeiro de 2020.

Cruz, C. F., Ferreira, A. C. S., Silva, L. M., & Macedo, M. A. S. (2012). Transparência da gestão pública municipal: Um estudo a partir dos portais eletrônicos dos maiores municípios brasileiros. *Revista De Administração Publica, 46*(1), 153–176.

Cuadrado-Ballesteros, B. (2014). The impact of functional decentralization and externalization on local government transparency. *Government Information Quarterly, 31*(2), 265–277.

Decreto-Lei n.º 89/2017 do Ministério das Finanças de 28 de julho. Publicada em Diário da República n.º 145/2017, Série I de 2017–07–28.

Deegan, C. (2002). Introduction: The legitimising effect of social and environmental disclosures—a theoretical foundation. *Accounting, Auditing & Accountability Journal, 15*(3), 282–311. https://doi.org/10.1108/09513570210435852.ISSN:0951-3574

European Parliament (2000). *Conselho Europeu de Lisboa 23 e 24 de março de 2000: Conclusões da Presidência.* Disponível em http://www.europarl.europa.eu/summits/lis1_pt.htm, consultado em 5 de outubro de 2019.

Eccles, R. G., & Krzus, M. P. (2014). *The Integrated Reporting Movement: Meaning, Momentum, Motives, and Materiality.* John Wiley & Sons.

Esteller-Moré, A., & Otero, J. P. (2012). Fiscal Transparency: (Why) does your local government respond? *Public Management Review, 14*(8), 1153–1173. ISSN: 1471–9045.

Ettredge, M., Richardson, V. J., & Scholz, S. (2001). The presentation of financial information on corporate Web sites. *International Journal of Accounting Information Systems, 2*(3), 149–168.

Eugénio, T. (2010). Avanços na Divulgação de Informação Social e Ambiental pelas Empresas e a Teoria da Legitimidade. *Revista Universo Contábil, 6*(1), 102–118. https://doi.org/10.4270/ruc.2010106

Franschini, A., Pereira, B., Duarte, C., Levy, D., Damke, E., Belizário, F., Santos, F., Bruno, G., Torres, J., Peixoto, M., Araújo, M., & Souza, Q. (2005). *Responsabilidade Social das Empresas: A contribuição das universidades. 4.* São Paulo: Peirópolis. ISBN: 978–85–7596–285–5.

Frey, M. R., Marcuzzo, J. L., & Oliveira, C. (2008). O Balanço Social como Ferramenta de Transparência para o Setor Público Municipal. *Revista Universo Contábil, 4*(2), 75–92. https://doi.org/10.4270/ruc.20084.ISSN1809-3337

Frost, G. R., & Seamer, M. (2002). Adoption of environmental reporting and management practices: An analysis of new south wales public sector entities. *Financial Accountability & Management, 18*(2), 103–127. https://doi.org/10.1111/1468-0408.00147

Fung, A. (2013). Infotopia: Unleashing the democratic power of transparency. *Politics & Society, 41*(2), 183–212. https://doi.org/10.1177/0032329213483107

García, A. C., & García, J. G. (2008). Determinantes de la divulgación de información contable a través de Internet por parte de los gobiernos locales: Determinants of Internet financial disclosure by local governments. *Revista Española De Financiación y Contabilidad, 37*(137), 63–84. https://doi.org/10.1080/02102412

García-Sánchez, I. M., Frías-Aceituno, J. V., & Rodríguez-Domínguez, L. (2013). Determinants of corporate social disclosure in Spanish local governments. *Journal of Cleaner Production, 39*, 60–72.

GRI. (2005). *Sector Supplement for Public Agencies: Pilot Version 1.0.* GRI. Disponível em https://www.globalreporting.org/, consultado em 8 de janeiro de 2020.

GRI. (2010). *GRI Reporting in Government Agencies.* Global Reporting Initiative. ISBN: 978–90–8866–033–7.

GRI. (2015). *G4 - Diretrizes para Relato de Sustentabilidade: Princípios para Relato e Conteúdos Padrão.* Global Reporting Initiative.

GRI. (2020). Global Reporting Initiative. Disponível em https://www.globalreporting.org/, consultado a 2 de janeiro de 2020.

Guillamón, M. D., Bastida, F., & Benito, B. (2011a). The determinants of local government's financial transparency. *Local Government Studies, 37*(4), 391–406. https://doi.org/10.1080/03003930.2011.588704

Guillamón, M. D., Ríos, A. M., & Vicente, C. (2011b). Transparencia financiera de los municipios españoles. *Utilidad y Factores Relacionados. Auditoría Pública, 55*, 109–116.

Guthrie, J., & Farneti, F. (2008). GRI Sustainability Reporting by Australian Public Sector Organizations. *Public Money and Management, 28*(6), 361–366. https://doi.org/10.1111/j.1467-9302.2008.00670.x.ISSN:1467-9302

Harduth, N., & Sampson, L. (2016). *A Review of the King IV Report on Corporate Governance.* Johannesburg: Werksmans Attorneys. Disponível em https://www.werksmans.com/wp-content/uploads/2013/05/061741-WERKSMANS-king-iv-booklet.pdf, consultado em 14 de janeiro de 2020.

Hollyer, J. R., Rosendorff, B. P., & Vreeland, J. R. (2011). Democracy and transparency. *Journal of Politics, 73*(4), 1191–1205.

IoDSA. (2016). *King IV: Report on Corporate Governance for South Africa 2016.* Institute of Directors in Southern Africa.

IoDSA. (2020). *King Report on Corporate Governance in SA.* Institute of Directors in Southern Africa (IoDSA). Disponível em https://www.iodsa.co.za/page/kingIII, consultado em 16 de janeiro de 2020.

Jensen, M. C., & Meckling, W. H. (1976). Theory of the firm: Managerial behavior, agency costs and ownership structure. *Journal of Financial Economics, 3*(4), 305–360. https://doi.org/10.1016/0304-405X(76)90026-X

Jorge, S., Sá, P. M., Pattaro, A., Lourenço, R. (2011). Local Government financial transparency in Portugal and Italy: A comparative exploratory study on its determinants. 13th Biennial

CIGAR Conference, Bridging Public Sector and Non-Profit Sector Accounting, pp. 9–10. Ghent, Belgium.

Joseph, C. (2010). *Sustainability Reporting on Malaysian Local Authority Websites* Curtin University of Technology, Australia. https://espace.curtin.edu.au/bitstream/handle/20.500.11937/2286/149374_Joseph2010.pdf?sequence=2&isAllowed=y, consultado em 13 de dezembro de 2019.

Joseph, C., & Taplin, R. (2011). The measurement of sustainability disclosure: Abundance versus occurrence. *Accounting Forum, 35*(1), 19–31. https://doi.org/10.1016/j.accfor.2010.11.002.ISSN:1467-6303

KPMG. (2017). *The KPMG Survey of Corporate Responsibility Reporting 2017*. KPMG.

Lodhia, S. (2010). Sustainability Reporting by the Australian Commonwealth Public Sector. *Proceedings of the 6th Asia Pacific Interdisciplinary Research in Accounting Conference (APIRA)*. Sydney, Austrália.

Marcuccio, M., & Steccolini, I. (2005). Social and environmental reporting in local authorities: A new Italian fashion? *Public Management Review, 7*(2), 155–176. https://doi.org/10.1080/14719030500090444

Mar-Molinero, C., & Serrano-Cinca, C. (2001). Bank failure: A multidimensional scaling approach. *The European Journal of Finance, 7*, 165–183. ISSN: 1466–4364.

Mussari, R., & Monfardini, P. (2010). Practices of social reporting in public sector and non-profit organizations: An Italian perspective. *Public Management Review, 12*(4), 487–492. https://doi.org/10.1080/14719037.2010.496262

Navarro Galera, A., Ortiz Rodríguez, D., Alcaraz Quiles, F. J., & Zafra Gómez, J. L. (2011). La divulgación de información sobre sostenibilidad en los gobiernos regionales y sus factores influyentes: el caso de España. *XVI Congresso AECA*, 1–24.

Navarro, A., Alcaraz, F. J., & Ortiz, D. (2010). La Divulgación de Información Sobre Responsabilidad Corporativa en Administraciones Públicas: Un Estudio Empírico en Gobiernos Locales. *Revista De Contabilidad-Spanish Accounting Review, 13*(2), 285–314.

Navarro Galera, A., Rodríguez, O. D., & Hernández, A. M. L. (2008). Identifying barriers to the application of standardized performance indicators in local government. *Public Management Review, 10*(2), 241–262.

Neto, J. M. F. T. (2019). Uma revisão da Teoria dos Stakeholders e principais pontos de controvérsias. *Revista Brasileira de Administração Científica, 10*(2), 1–16. Disponível em https://doi.org/10.6008/CBPC2179-684X.2019.002.0001, consultado em 10 de dezembro de 2019.1–61. doi: https://doi.org/10.6008/CBPC2179-684X.2019.002.0001. ISSN.:2179–684X

Nevado-Gil, M. T., & Gallardo-Vázquez, D. (2016). Información sobre Responsabilidad Social contenida en las páginas webs de los ayuntamientos. Estudio en la región del Alentejo. *Revista Española de Documentación Científica, 39*(4), 1–17.

Piotrowski, S. J., & Van Ryzin, G. G. (2007). Citizen attitudes toward transparency in local government. *The American Review of Public Administration, 37*(3), 306–323.

Prado, J. M., García, I. M., & Cuadrado, B. (2010). Sustainable cities: Do political factors determine the quality of life? *WIT Transactions on Ecology and the Environment¸142*, 39–50.

Ribeiro, V. P. L., & Guzmán, C. A. (2012). La Gestión Medioambiental en las Entidades Locales Portuguesas. *Revista Iberoamericana de Contabilidad de Gestión (RIGC), X*(19) 1–30.

Ribeiro, V. P. L., & Guzmán, C. A. (2011). La Divulgación de Información Medioambiental en las Autarquias Portuguesas: Caracterización y Fatores explicativos. *Revista De Contabilidade e Gestão - Revista Científica Da Ordem Dos Técnicos Oficiais De Contas, 11*, 43–74.

Ribeiro, V., Monteiro, S., & Moura, A. (2018). Divulgação de Informação na Internet sobre Responsabilidade Social—Evidência Empírica nos Municípios Portugueses. *Revista De Gestão Ambiental e Sustentabilidade, 7*(2), 185–209. https://doi.org/10.5585/geas.v7i2.373

Río, O. A. V. C., & Toharia, L. (2009). Un Análisis Espacial del Desempleo por Municipios. *Revista de Economía Aplicada, XVII*(49), 47–80. ISSN: 1133–455X.

Sciulli, N. (2009). Sustainability reporting by local councils in Coastal Regions: Na Australian Study. *Asian Journal of Finance & Accounting, 1*(1), 76–86. ISSN: 1946–052X.

Serrano-Cinca, C., Rueda-Tomás, M., & Portillo-Tarragona, P. (2009a). Factors Influencing E-Disclosure in Local Public Administrations. *Environment and Planning C: Government and Policy, 27*(2), 355–378.

Serrano-Cinca, C., Rueda-Tomás, M., & Portillo-Tarragona, P. (2009b). Determinants of e-government extension. *Online Information Review, 33*(3), 476–498. ISSN: 1468–4527.

Slack, R., & Shrives, P. J. (2008). Social disclosure and legitimacy in Premier League football clubs: The first ten years. *Journal of Applied Accounting Research, 9*(1), 17–28. https://doi.org/10.1108/09675420810886105

Smith, K. A. (2004). Voluntarily reporting performance measures to the public: A test of accounting reports from U.S.Cities. *International Public Management Journal, 7*(1), 19–48. ISSN:1096–7494.

Sol, D. A. (2013). The institutional, economic and social determinants of local government transparency. *Journal of Economic Policy Reform, 16*(1), 90–107. https://doi.org/10.1080/17487870.2012.759422.ISSN:1748-7889

Tilling, M. V. (2004). Communication at the Edge: Voluntary social and environmental reporting in the annual report of a legitimacy threatened corporation. *4th Asia Pacific Interdisciplinary Research in Accounting (APIRA) Conference,* 1–27. https://researchonline.nd.edu.au/bus_conference/7, consultado em 11 de dezembro de 2019.

Tilt, C. A. (2001). *Environmental disclosure by Australian companies: What is happening outside the annual report?* University of Adelaide. Disponível em http://citeseerx.ist.psu.edu/viewdoc/download?doi=10.1.1.196.4160&rep=rep1&type=pdf, consultado em 20 de novembro de 2019. ISSN: 1441–3906.

Transparency Internacional Portugal. (2020). Índice de Transparência Municipal. Disponível em https://transparencia.pt/itm/, consultado em 11 de dezembro de 2020.

Chapter 9
School Citizen Assemblies: Developing Educational Ecosystems of Civic Engagement, Action and Social Change

Chris McLean

Abstract School Citizens Assemblies (SCA) seek to empower young people as changemakers by assembling schools, experts and community stakeholders through a collaborative and co-creative process of research, innovation and action. Understanding and solving complex problems requires people, organisations and governments to work together in creative, informed, inclusive, sustainable, responsible and compassionate ways. By working collaboratively, we can appreciate the complexity, diversity and difference that underlie these challenges and mobilise our knowledge and skills to create effective and sustainable actions and solutions. During this process of analysis and challenge-led learning, it is also important to unpack the different relational assemblages and black boxing that can limit or restrict an inclusive and equitable process of action and interaction (Latour, reassembling the social: An introduction to actor-network-theory, Oxford University Press, 2005). The SCA process seeks to achieve this in two ways. First, by increasing divergent thinking and opening spaces to unpack creativity, diversity and difference (for example, by viewing 'matters of fact' as 'matters of concern'). Second, by creating spaces for convergence that enable effective actions to be created, tested and evaluated through real-world learning. Our educational system requires additional approaches that can support innovative and creative approaches to teaching, learning and community engagement. SCAs seek to achieve this by enabling different stakeholders to work together to tackle complex local and global problems and providing the conditions for individuals, organisations and the planet to flourish and thrive.

Keywords School citizen assemblies · Civic engagement · Climate change · Learning · Actor network theory · Sustainability · Social action · Social responsibility · Education

C. McLean (✉)
University of Manchester, Manchester, UK

D. Crowther and S. Seifi (eds.), *Preparing for a Sustainable Future*,
Approaches to Global Sustainability, Markets, and Governance,
https://doi.org/10.1007/978-981-99-2456-1_9

9.1 Introduction

Imagine a room full of young people discussing the future of our planet—a vibrant and thought-provoking discussion spanning different issues and invoking a sense of agency and belief that together they can make a difference. Not only in their local community, but also issues at a global scale. These young people are exploring a specific challenge and are gathering research evidence from a variety of sources. They have already begun to assemble subject-specific knowledge from different disciplines aligned with the school curriculum and through desk-based research. Experts have also provided talks and resources to help them understand the issues in greater detail and some projects involve the pupils and local organisations working together on specific projects.

These young people have also engaged in a research process that focused on them understanding the problem from different perspectives, actively listening to different points of views and developing a sense of empathy and compassion for the different stakeholders. Experts and stakeholders have highlighted a range of different ideas, issues, problems and solutions, alongside their experiences, concerns and hopes for the future. While all the different views are respected and considered through the process of analysis, they are also critically examined in the context of the challenge. This is key and the different activities connected with the process encourage critical thinking, collaboration, compassion and creativity. In addition, these activities help those involved to develop a greater understanding of different points of view, the process of redefining problems, creating new ideas and solutions, and the importance of engaging in an iterative process of reflection, review and refinement. As well as advancing a deeper knowledge and understanding of the issues, this space creates new opportunities to enable young people, organisations and communities to take positive action to improve the world around them. The space is a School Citizen Assembly.

School Citizen Assemblies (SCA) create civic and educational ecosystems that enable young people to work with experts and stakeholders to enact change and positive social action in their local and global communities. Schools act as hubs bringing together different experts and stakeholders through the SCA process. This can include experts in universities, colleges, businesses, charities, public and third sector, as well as the wider stakeholder voices within the community. The aim of the SCA is also to develop partnerships between different people, groups and organisations seeking to solve a specific challenge. This is an ecosystem of young people, businesses, organisations and stakeholders working together to develop a greater understanding of complex problems underlying the challenge and effective and sustainable solutions. A process that supports them to work together to make a positive difference to many people and world around them (Nichols & Cator, 2008).

For young people, these spaces also enable them to amass a wealth of knowledge, skills and experiences by engaging with experts and organisations connected to the challenge. By focusing challenges and actions on real-world issues and complex problems, young people can also feel more passionate about tackling these issues and develop a greater sense of agency, empathy, and connection with their local and global communities. While SCAs are excellent for learning about and tackling problems collaboratively, such as climate change, this process can also work well for a whole range of different issues and problems that a school and community may wish to examine in more depth. For example, these could include issues highlighted within the UN Sustainable Development Goals (UN, 2016).

Stakeholders also go beyond the people and organisations within our communities. We also need to work closely with nature and the world around us when developing productive alliances to ensure rich and sustainable futures for all. This will require the implementation of effective solutions and positive impacts to tackle the major challenges that face us now and, in the future, (climate change, biodiversity, economic crisis, conflicts). By broadening out the voices and alliances in this way (to nature and the world beyond a purely human focus), it becomes possible to engage differently with and understand the problems, concerns and solutions from many different perspectives and concerns (Latour, 2005, 2013). Also, thinking through complex problems and possible solutions in this way places a greater focus on active listening and compassion for different viewpoints and from a wider perspective.

Underlying the SCA approach is a rich and diverse set of practices, research and ideas connected to the concept of Design Thinking (DT). During the early stages of development, while some academics expressed an interest in developing and researching the DT approach, there was much more of a practitioner focus. However, there has been an increasing interest in Design Thinking from many different interdisciplinary domains (Lor, 2017; Melles et al., 2015). This includes a growing body of research conducted exploring the role of this approach in many different areas of study and more specifically connected to organisations and management practice.

Within this chapter we will examine the important interconnections underlying the design, development and application of the SCA approach into schools and communities and how it links to different conceptual thinking relating to the ideas of design and ecosystems of learning and social action. This will include examining different thinking and approaches connected to this approach and the ways in which this process seeks to rethink action and social change, through community action and education.

9.2 School Citizen Assemblies: Designing and Developing SCAs

Designing and developing the SCA process has involved bringing together a broad range of partners and organisations. Many different experts and organisations are contributing to this process and in different ways. Schools and teachers are helping to design the process, along with experts in Challenge Led Learning and Design Thinking (Lor, 2017, Melles et al., 2015), Dilemma Led Learning (Kidd & Roberts, 2018), Problem-Based Learning (Leat, 2017), creativity, assessment and curriculum design, citizen assemblies, community action, sustainability and climate action.

Public sector, third sector and charitable organisations, universities and colleges, business and other organisations are also all working together on this project to help take forward this design process. Together we are designing, developing and implementing SCAs in the form of toolkits, training, teaching and learning resources (videos, documents, guides, toolkits, activities, programmes and different example formats) and a 'Meet the Experts' directory. These will be tested and developed through pilots and refined through this ongoing process of co-creation. The 'Meet the Experts' directory has been particularly well received. It will provide a detailed search engine that will enable schools to easily connect with experts in their community and open the SCA process to a range of voices aligned with a specific challenge, issue or themes. This will enable experts, young people, schools and stakeholders to work together to enact change and make a positive difference both locally and globally.

Currently we are engaged in pilot projects, working with teachers, schools and expert groups to design, develop and implement SCAs within primary and secondary schools. This action-oriented research and practice seeks to build partnerships between an educational ecosystem of universities, colleges and schools. In particular, the SCA approach is based on schools as hubs linking communities, experts, teachers, pupils and other organisations around a specific issue and design challenge. The initial SCA project focuses on climate action. However, the SCA process could examine a wide range of local and place-based issues. These could include issues connected to equality, inclusivity, loneliness, mental health, drug use, crime, intergenerational support, foodbanks, pollution, waste, social care, to name but a few. A community may also have very specific place-based concerns that they wish to focus on and SCA provides an effective approach to open-up for discussion, debate and action around these issues.

By aligning the conceptual thinking relating to this project with the pedagogical and collaborative design and democratic deliberative assembly approaches, we also wish to highlight the shared interests of different individuals, groups and institutions relating to learning, education, community action and climate change. In addition to emphasising the potential benefit of this approach to different people, organisations and society, we also wish to avoid naïve assumptions regarding agency, choice and action (Aroles & McLean, 2016; Cooren, 2018; Latour, 2013; Tsoukas & Chia, 2002). For example, assuming a simple and essential process of free and informed choice, in which you merely open-up spaces where people define 'rational' problems and act on these based on a simple and linear view of decision-making (McLean & Aroles, 2016). In contrast, we wish to acknowledge the tensions, controversies and issues connected to the process. For example, how issues emerge as problems and how they become connected to solutions. This includes exploring how certain truths become central to a discussion and others become rejected; which voices are listened to and why; how certain paths may be taken when defining problems and others are rejected; and how we encourage spaces for creativity, innovation and experimentation, while also acknowledging the importance of thinking through actions, outcomes and impact (Latour, 2013).

We acknowledge that it is not possible to simply uncover all the tensions that underlie a design challenge and fully 'know' how different outcomes and pathways should emerge. However, SCAs can provide greater opportunities and spaces for different voices and outcomes to emerge, develop and promote positive outcomes for people and the world around us and support collaborative action. In doing so, the SCA process seeks to educate young people about the process of crucial thinking that includes the ability to understand and empathise with different perspectives, examine the quality and credibility of different sources of information and truths and develop solutions through co-creative and collaborative approaches. For instance, the aim to empower young people as change makers includes developing a greater capacity to: listen, understand and empathise with difference, diversity and complexity; develop higher order thinking skills (critical thinking; problem solving and analysis, creativity, imagination etc.); collaborate through teamworking and a process of co-creation; develop greater resilience, strength of character and positive values, become flexible and adaptable thinkers, and to have a greater connection with their communities. The SCA approach also values both the importance of disciplinary knowledge and skills and interdisciplinary learning, so that we can mobilise knowledge, skills and values through real-world learning and practice.

Consequently, SCAs seek to empower young people and communities as change-makers to tackle local and global issues by creating spaces where communities can work together to act collaboratively, and schools can act as hubs to support this process. Not only do they provide communities with the capacity to respond to the challenges that face us now and, in the future, they can also help to provide young people with a greater sense of agency and voice; compassion and understanding of difference; and the knowledge, skills and understanding to delve deeper into solving complex problems. Through this process, SCAs therefore seek to unite the ability to mobilise knowledge and skills through the key competencies of compassion, collaboration, creativity and critical thinking.

Within the next section we will examine some of the relevant conceptual issues underlying the SCA aligned to Actor–Network Theory, before moving onto a review of the design, development and implementation of the SCA through the interweaving of these different ideas, approaches and methodologies.

9.3 School Citizen Assemblies: Enacting SCAs Through the Lens of Actor–Network Theory

There are many different connections between the ideas and thinking underlying this chapter and a broad range of literature and areas of study. This includes discussions and debate concerning the role of citizen assemblies within society; ideas associated with communities of practice; creative design thinking; environmental education, wider pedagogical issues concerning school curriculum design and assessment; community action and change; the role of civic universities; real-world and problem-based learning; organisational design and change; sustainability and climate action, the development of global and UN SDGs; and the conceptual thinking connected to Actor–Network Theory (ANT) (Akrich, 1992; Latour, 2013; Law, 1992). There is neither the time or space to go into detail about the many and diverse connections to these different ideas and approaches. However, one key conceptual connection that will be examined within this chapter and in relation to the design, development and initial implementation of the SCA approach, is the Actor–Network Theory approach.

In seeking to reject taken for granted and essentialist assumptions about reality (i.e. the discrete, independent and a priori existence of people and things), Actor–Network Theory seeks to dissolve and examine well-established divides (e.g. nature/culture, structure/agency, macro/micro, etc.). This includes examining the distributed action and agency that underlies the emergence, translation and stabilisation of specific outcomes through everyday practice. ANT seeks to focus on the emergence and translation of heterogeneous networks of actors and the various negotiations, dynamics and trials of strength that underlie this process (Latour, 2013). For example, how different forms and types of alternative energy sources are emerging as rivals to fossil fuels; how some truths and facts become black boxed and stabilised through different sets of relations, while other pathways are rejected or fade away. Rather than assuming a neutral and value-free process, ANT can help to unpack the values, ethics and power relations embedded within this process. Through this shift in thinking, it becomes possible to study the complexities and difference that underlies the assemblage and translation of heterogeneous material and non-material interests, forces and relations (Aroles & McLean, 2019; Marble, 2012). This conceptual thinking has been key to the emergence of the SCA and the thinking underlying this approach.

Another key feature of ANT is shifting our attention away from focusing merely on people as key stakeholders and agents in the change process, to the many other actors connected to change and action. Furthermore, rather than viewing interests, needs or views as discretely and simply located in people, artefacts, techniques or practices, ANT shifts our attention to the emergence of interests and needs through complex relational and distributed processes (see Akrich, 1992). ANT therefore seeks to unpack the complex dynamics and relations in which interests continually align/disalign, emerge and disappear, are mobilised or excluded, and how this process impacts on the emergence of outcomes and unintended consequences (Fenwick, 2011; Gherardi & Nicolini, 2005; Singleton & Law, 2005; Venturini, 2010).

Additionally, in contrast to a simple and linear process of cause and effect, ANT examines the iterative process of becoming, assemblage and translation in which practices, relations and outcomes emerge through a continual process. This enables a rethinking and unpacking of ideas such as relationality, logics of action, distributed agency and heterogeneity. Making sense of the dynamics and complexities of translation, stabilisation and the different trials of strength also requires the ability to examine the continual work and relational effort underlying this process (Cooren, 2018). This includes unpacking situations of convergence around specific 'matters of fact' (those that may become taken for granted or black boxed) and opening spaces to consider 'matters of concern', difference and the making of 'facts'. This requires an ability to engage in divergent thinking and critical analysis in order to examine the complexity, multiplicity and difference in more depth (Latour, 2013; Moberg, 2018). As we will see later, these are also key themes underlying the pedagogies and approaches connected to the SCA process. For example, the application of challenge-led learning, design thinking and dilemma-led learning connected to divergent and convergent processes (Leat et al., 2012).

By examining the complex assemblage, alignment and translation of different interests connected to specific relational encounters, we can begin to appreciate how interests emerge and others fade away within specific settings leading to particular outcomes and sometimes unintended consequences (Aroles & McLean, 2016; Latour, 2013). Gaining greater insights into the process of translation involves attending to two specific aspects. Firstly, developing a clearer understanding of the emergence and dynamics that underlie this process (especially with competing interests and scripts), and secondly, examining the emergence of specific relations and outcomes through this process (subjects, objects, boundaries and divides). This then provides a greater insight and understanding of the multifaceted dynamics, forces, actions and potentialities that surround everyday life (Aroles & McLean, 2016). For example, the importance of pedagogical practices of learning and community action that create more spaces for critical thinking, collaboration, compassion and creativity. This includes creating spaces that enable different people and organisations to raise different points of views, concerns and questions around specific problems and solutions and the ability to hear and listen to a diverse range of perspectives and possibilities connected to a specific issue or challenge.

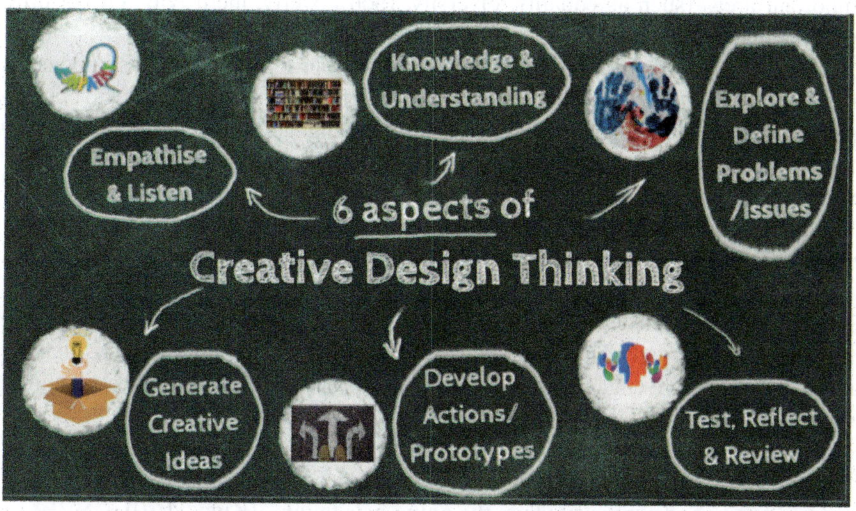

By drawing on different approaches and pedagogies connected to challenge-led learning, design thinking and dilemma-led learning (creativity, problem-based learning, oracy and deeper learning), we can begin to see the value of these approaches to learning and collaborative action. These approaches highlight the importance of reflecting on: complexity and diversity; the dynamics of convergent and divergent (disruptive) thinking; and the emergence of different truths, problem and/or solutions within specific settings. For example, this includes the importance of listening to different voices and ideas, how certain voices and forms of actions may be devalued or excluded, and issues concerning the processes of convergence or coherence that may value interests and voices differently depending on the context. This also raises the questions, such as: How can the SCA process enable different challenges and matters of concern to be explored from different perspectives? How can this process create spaces for divergence and convergence that enables greater levels of creativity, critical thinking, compassion and collaboration between young people, experts, stakeholders, organisations and communities?

9.4 School Citizen Assemblies as a Force for Good?

School Citizen Assemblies provide a way of connecting schools, communities and experts from educational institutions, businesses, charities, third and public sector organisations to explore and tackle many different local and complex issues. With schools acting as the hubs, they can assemble different agencies and create spaces where pupils, teachers, experts and communities can connect around a specific problem or challenges. In addition to tackling significant issues such as climate action, SCAs can be an effective tool to make a difference in a wide range of other important areas (such as equality, diversity, food poverty, inclusivity, crime, drug use, loneliness, etc.).

By generating creative and innovative spaces, young people, experts and stakeholders can work together to tackle complex problems (through divergent and convergent thinking), while also encouraging continual engagement with different viewpoints, tensions and obstacles that may prevent effective change or create problematic unintended consequences. Part of the methodology for achieving this thinking is through challenge-led learning. This includes an iterative process and activities that enable divergent and convergent thinking.

While this approach has been developed from broader ideas of design thinking, the SCA process has been developed and adapted it to fit with teaching and learning practice within schools and approaches such as, Dilemma-Led Learning. This has enabled us to produce a more engaging and productive approach for learning and community engagement. For instance, Dilemma-led learning (DLL) encourages young people to explore "*some deeply philosophical questions and grapple with what it is to be human...That is the essence of good education—working at the edges of morality and figuring out where we sit when the going gets tough.*" (Kidd, 2020).

Consequently, SCAs not only provide an engaging process of learning for young people to develop a range of knowledge, skills, values and capacities (learning outcomes), they also seek to create inclusive processes for community action. This can be particularly effective for young people who are less engaged or find traditional approaches more difficult (SEND, disadvantaged groups, pupils who disengaged with learning). Beyond SCAs benefiting young people, they can also create productive outcomes for schools, organisations, communities and society in general by:

- Connecting and aligning different aspects of the curriculum through an interdisciplinary approach.

- Mobilising knowledge and higher order thinking skills through real-world learning.
- Improving pupil learning outcomes in a wide range of subjects and areas.
- Developing partnerships between schools, experts and local community members to tackle complex and local issues and problems.
- Enriching the teaching experience by embedding this learning process into the curriculum and the overall learning aims of the school.
- Empowering young people by creating a greater sense of agency and voice.
- Providing more opportunities for young people to make a real difference in their local and global communities and empowering communities to act and create positive change.
- Preparing young people for their future careers and lives.
- Developing young people with a range of knowledge, skills, values and understanding that will be important for the long-term sustainability of organisations and society.
- Connecting young people to their local communities through greater engagement, agency and sense of belonging. This includes working with others to make a difference and create new ways of tackling local problems and creating new opportunities.
- Position schools as hubs to create spaces for collaborative action and social change.

These wide range of learning outcomes and competencies connected with SCAs also align well with many different national and international initiatives (UN, 2016; UNICEF, 2019; UNSECO (Wals, 2012); OCED, 2018).

9.5 How Can SCAs Enhance Young Peoples' Knowledge, Skills and Understanding?

Developing young peoples' knowledge, skills, compassion and understanding are key to the SCA process. SCAs can therefore provide effective forums to enhance learning experiences by mobilising knowledge and skills and enabling young people to reflect on different positions and viewpoints. This can be particularly effective if the SCA process is built into the school's curriculum and there is a cross-fertilisation between SCA knowledge development, the learning process and curriculum design. Beyond the focus on developing knowledge and skills, SCAs are also 'values rich' as they support schools in developing curriculums that seek '*to nurture in young people, the ethos and ability to care for oneself, for one another and the natural environment, for present and future generations*' (Jim Knight, 2021). Pupils are also encouraged to challenge and explore in greater detail specific views, positions and thinking around certain themes and connect this to their understanding of different values and ways of viewing problems and solutions.

SCAs help to provide the structure to engage in this form of critical thinking as young people are introduced to a range of activities connected to a specific challenge.

These encourage them to explore the complex and diverse viewpoints that underlie certain problems and/or solutions. In the words of Latour (2013) they open spaces and unpack controversies and 'matters of fact' by reimagining them as 'matters of concern' (Latour, 2013). It also puts young people in a better position to examine how a wide range of knowledge, understanding and skills can produce different ways of viewing a problem and understanding an issue. For example, it cultivates the pupil's ability to hear and listen to multiple viewpoints, explore complexity, diversity and difference, and the capacity to tackle different challenges and identify opportunities. Through these critical thinking activities pupils are more likely to develop the knowledge, skills, values and capabilities to produce a future world in which we can all thrive together more effectively and sustainably in alignment with our planet. Through the ideas of distributed action and agency underlying ANT young people are also encouraged to rethink stakeholders beyond humans and understand the importance of considering the interests of animals, objects and nature when exploring challenges and developing solutions. Thinking through problems and developing solutions requires an ability to unpack the complexities of distributed action and agency in order to develop effective and impactful solutions. Furthermore, SCAs encourage collaboration, teamwork and a process of co-creation to ensure that we work together to understand problems and develop solutions in divergent and convergent ways.

The image below highlights the iterative process aligned to the SCA. While some schools may decide to engage in the process over a single day, others may break it down over several sessions and embed it more into the school curriculum over a term or school year.

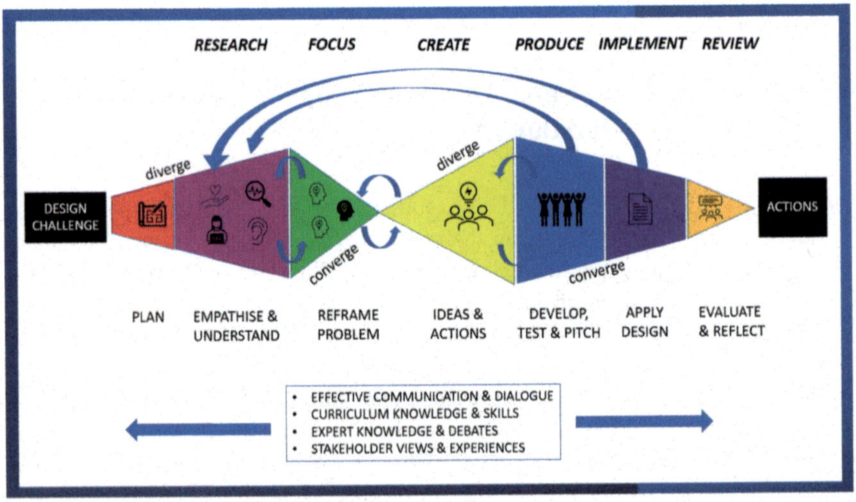

This could involve breaking down the design challenge process over a series of weeks or developing some prior sessions to help to develop the pupil engagement, skills and understanding of the process.

SCAs can also achieve this by combining a range of different pedagogies, methodologies, activities and practices. These include Dilemma Led Learning, Creative Design Thinking, Problem-Based Learning, creativity, and ideas connected to the citizen assemblies approach. It also seeks to connect this process to other issues associated with the management and organisation of schools and teaching and learning practice. These include curriculum design, the leadership and management of schools, disciplinary knowledge and interdisciplinary learning, as well as real-world learning aligned with local and global communities.

9.6 Making SCAs Work and Embedding Them into Schools

The design and implementation of SCAs also require the ability to appreciate the relationship between divergence and convergence. While SCAs need to work at multiple levels and in different forms depending on the school, a set of clear, distinct and well-designed elements (systems, toolkits, resources, platforms) are required that provide some structure and coherence to the process of design and implementation. The design and development process may differ depending on the development of a SCA approach within a single school (school-wide, a specific key stage, year or group), a Multi Academy Trust, different schools in a region and with other schools internationally. During the SCA, there are also different ways of linking the process to the local community (in terms of experts, user groups, participants, etc.).

The shift between divergent and convergent practices and activities can also be beneficial to students with different learning and engagement styles and degrees of confidence. While some students appear to be engaged with more expansive and creative thinking connected with divergent activities, others seem more comfortable with more convergent activities. While activities can be organised to enable students to benefit from their current tendencies, they can also help them to push their boundaries and develop new skills. SCAs also build in activities that enable students to think individually as well as in teams. Again, this supports a range of different styles of engagement and thinking (for example, a more introverted learner may benefit from more time to introspect, while other students may enjoy more share-storming where teams work together to solve problems). The aim within the SCA is to provide spaces for both and a balance of activities that reflect different styles and approaches to learning and action.

Another option for schools is to start small and build to larger forms of SCAs as the process develops (one-day workshops, several sessions over a term or year, focusing on different issues). SCAs can also work at different key stages (ages) and can be tailored to suit certain age groups, subject areas or challenges. For example, a school may wish to use a SCA approach to help with primary to secondary transition (Year 6 to 7). The SCA process could also be extended to connect with university students who could engage with schools as experts and/or participants in the development of community actions. Another option when designing a SCA is providing greater student involvement in the process of design. While the school and teachers may select the overall focus such as climate change and biodiversity, they may allow the students to decide on the specific challenge they wish to focus on for their projects.

Maintaining a balance between the dynamics of divergence and convergence is also key to the management and introduction of this approach within the school. The approach needs to be clear, simple and straightforward, particularly in the early stages of introduction. However, it also needs to be designed in such a way to allow flexibility and adaptability to the local conditions of the school, location and the interests of the teachers and young people. Additionally, a SCA needs to capture the complexity and diversity connected to specific challenges and issues, while also enabling a clear and effective process of design that connects with school curriculums and the realistic and achievable outcomes. The SCA may be seen as less successful if it is too restrictive or rigidly defined, or too loose or complicated in terms of the process or outcomes. Therefore, it is important to design it in such a way that the school and community can balance the dynamics of divergence and convergence, clarity and complexity.

SCA activities also need to be built into the process that supports divergence and convergence. Divergence to support creativity, new thinking, imagination, difference and new lines of action (unpacking matters of fact and creating new ideas, opportunities and innovative pathways). Convergence to support the definition of problems and the enactment of new ideas actions, prototypes and different solutions (translating interests in black boxes and punctionalised actors). Other forms of convergence and matters of fact also need unpacking and reflecting on when evaluating the design and development of the SCA process. These include issues around accountability and assessment issues; timetabling; curriculum design constraints; subject and disciplinary divisions; availability of time and effort within schools; access to experts; and connecting up stakeholders in the community; to name but a few. Thinking through these issues are also significant when considering the potential barriers to learning and introducing new methodologies, pedagogies and curriculum design. Therefore, considerations of divergent and convergent thinking is not only relevant to the application of SCA in practice, but also the process of designing an SCA within a school environment. This includes approaches to assessment, accountability, leadership and organisational change.

9.7 Potential Hurdles to Overcome: Assessment, Accountability and Organisational Change

While schools are regulated by many different rules, requirements and processes in terms of assessment and accountability, many teachers have expressed concerns with some of the current drivers of teaching and learning practice. This includes a broader concern within society that schools are motivated to act as exam factories focused on specific subjects and exam questions, rather than broader educational experiences. Part of this relates to a concern that some drivers place limitations on achieving the true purpose of education and the provision of effective learning experiences to ensure every pupil achieves their 'full' potential. Ensuring school leaders support the SCA process will also be key to the effective implementation of the approach. While there may be a group of teachers promoting and developing the approach within the school, it will also be important to build alliances and ensure a process of collaboration and inclusivity throughout the school. As Kidd and Roberts (2022) explain with reference to Dilemma Led Learning (DLL):

> *While many teachers seek to make learning engaging and exciting for children, an overloaded curriculum, pressure of testing, marking, administration and other tasks, it can be difficult to maintain this in the classroom. Teachers may question the idea of introducing more into a subject area of curriculum. The wonderful thing about DLL is it needn't be another subject. It's a cuckoo pedagogy that can be slotted into almost any area of learning and the benefits are enormous.*

It is also important to show how the SCA process can help to align shared interests and produce many positive outcomes from many different actors (pupils, teachers, communities, organisations and the natural world around us). Highlighting the flexibility and adaptability connected to SCAs is also important to ensure that teachers feel that they can retain their professional role and status when connecting SCAs to their own ideas and approaches to teaching and learning and community action.

9.8 Aligning Shared Interests and Opening Up Spaces for Reflection and Critique

The SCA process is one element within a broad set of strategies, pedagogies and practices of teaching and learning that a school may embed and connect with their overall curriculum design. SCAs also connect well to a wider vision of education and learning expressed by teachers and others beyond the school gates. The SCA concept has therefore been very popular when considering the many different interests and thinking around education. It also engages well with rethinking learning and assessment in ways that respond to the complexity and unpredictability that surrounds us and prepares young people for the complex future world that will face them. In addition to fulfilling the aim of achieving high pupil outcomes (from a traditional perspective), SCA also extends this to a wealth of further outcomes for both young people and society.

SCAs also seek to avoid reinforcing false dichotomies (such as those between knowledge-rich and problem/skills-based learning). It overcomes some of these tensions by valuing both disciplinary knowledge and interdisciplinary learning. It also seeks to mobilise knowledge and learning through real-world issues and interweaving these through the process. This provides greater opportunities to apply knowledge to specific situations and enable young people to develop a greater understanding and empathy with other views and issues. It also opens up spaces to question 'knowledge', 'facts' and 'truth' and challenge how different ideas connect to real-world problems and issues.

Due to the significant pressures schools and teachers face within the current system, the SCA seeks to support schools develop this process through high quality, adaptable and flexible resources. This will include CPD training and expert support connected to the issues (example programmes, toolkits, alternative formats, information and links to experts). For example, many teachers have raised concerns that either they do not feel they have the full range of knowledge or expertise to teach environmental education, or there is limited room in the current curriculum or teaching week to focus on this area. Therefore, the SCA process can start off small and grow organically so that teachers do not feel that it becomes an additional strain and duty to be performed.

Again, this aligns well with the ideas of ANT in terms of translating and aligning shared interests. For example, the SCA toolkit will include activities, example programmes, suggested formats, ideas to map to curriculum design and other aspects that could be linked to the more convergent view of education. However, the SCA process also seeks to open spaces for divergent thinking and new ideas. Therefore, the dynamics of divergence and convergence are not only key to how you design and apply the SCA within the school, but also how you select different activities that seek to encourage innovative approaches to tackle real-world problems.

Deciding how to connect and embed SCAs into curriculum design is therefore a key part of the process to link and mobilise knowledge and learning. It is also important to ensure that SCAs fit with the ways in which different schools design their curriculum allowing for the local conditions and the location of the school. Therefore, SCAs need to adapt and flex to these conditions. Teachers also require high-quality resources and training to ensure that they can effectively implement SCAs through different formats and programmes of activities that suit their school and setting.

9.9 Different SCAs Forms: Supporting Flexibility and Adaptability

Within a school, a SCA can range from simple version (a one-day workshop with presentations and insights from experts and local participants); to more sophisticated versions over a term or whole year, with greater levels of discussion and analysis. A selection of proposed formats, activities, case studies and toolkits (SCA methodologies and pedagogies) are being developed to support teachers running SCAs. This process enables schools and teachers to embed SCAs into their current and future thinking for curriculum design and select their own preferred activities to teaching and learning in order to make their own SCA model. A crucial aspect of making the SCA process work well is ensuring it connects well to teaching and learning within the school and provides a flexible and adaptable approach for the teachers to make connections to their curriculums.

Also, rather than a single workshop, SCA sessions could be spread over several weeks and embedded further into the school curriculum and real-world projects.

This enables the pupils to pitch and test different thinking, ideas and solutions with experts and stakeholders in order to gather feedback and suggestions to help with this iterative design process (e.g. asking individuals from the community into the school to review the actions). This aids the ongoing and iterative process of design that underlies challenge-led learning, as it allows the pupils to reflect on this feedback before finally implementing further actions. They can also critically reflect on what went well and what they would do differently with a similar future challenge through review and evaluation sessions. Overall, the SCA process enables pupils to learn about the benefits of regularly connecting back to expert and stakeholder voices in the community through different stages of developing design. This provides a richer knowledge, understanding and empathy with stakeholder views and enables the development of creative and innovative actions and solutions, to help tackle issues such as climate change and biodiversity.

SCA can also work between schools locally, with other schools in different regions and even with schools and/or communities in different countries. Another option is to engage to a greater extent with the local community and invite in local parents, businesses, organisations and other representatives from the community to provide different voices to aid the process of design. Engaging the local community in the process of developing actions is key to effective social change and linking with local projects connected to environmental issues can also be an effective mechanism for local action. This process provides an excellent way to support local change and young people can be empowered to play a role in creating new strategies and actions to support this change. Project work can also connect local change with global outcomes (for example, encouraging local people and businesses to purchase more sustainable palm oil or reducing carbon emissions).

Pupils can also expand their learning experiences by developing the knowledge and skills to organise and run a SCA themselves, in turn increasing their involvement in the changemaking process. The SCA process is also likely to make young people feel more engaged with their own learning and development and empower them as changemakers with the voice and agency to impact on social action within their local and global communities.

9.10 Future Thinking and Plans: Could SCAs Work in Other Areas Beyond Climate Change?

The SCA approach can be easily adapted to examine many different issues relevant to any school. While the initial SCA process and pilots will focus on environmental education and climate action and training and resources will attend to these aspects, future SCAs will focus on wider aspects. Further resources will be developed in other areas (equality, diversity and inclusivity, food/fuel poverty, social care, skills development, mental health, etc.) to support alternative forms of SCAs. One way of effectively framing these different issues is through the UN SDGs.

Developing partnerships with experts, communities and others who contribute to the knowledge, skills and understanding of the young people is also a key aspect of the SCA process. Schools will have access to a '*Meet the Experts*' directory. A search engine that will enable schools and teachers to find experts who can provide support in different ways. This could include links to resources and information, online Q&A sessions or schools visits (as part of the SCA). They will also be able to search by region so that they have access to local experts in their locality. There are also plans to engage university students in this process of engagement, so that they can provide expertise and support to schools and participate in SCAs.

Connecting to local problems and issues is also key to this process, as pupils can also learn about the importance of language and changemaking through hearing a plurality of views and by reflecting on the complexity of issues through real-world problems. This will not only impact on how they view problems and solutions, but also how they seek to enact these through SCA actions and produce effective influencing strategies. For example, showing how two narratives can be combined to inspire behaviour change ('brutal facts' of the climate crisis, alongside invoking a sense of hope for a more sustainable future).

A recognition of different age groups and approaches or experiences to learning is also required. SCA seeks to be inclusive and develop greater energy and passion for learning that will engage all pupils. However, to be inclusive and engaging it is important to provide activities that suit different ages and groups. The aim is to be aspirational and support pupils through this process so that they can all experience high-quality learning experiences and outcomes. One thing that has emerged through our previous research into dilemma-led learning (DLL) and real-world learning was the extremely positive response by teachers to this approach. This included the introduction of this alternative set of activities and pedagogies that had transformed the learning experiences of some pupils (e.g. SEND and disadvantaged pupils) from struggling and disengaged (using traditional approaches) to much more passionate and engaged with the learning process based on DLL and real-world learning. Not only did their learning outcomes in a wide range of areas improve, but also their attendance levels. This highlighted that the positive benefits of DLL went beyond certain activities, as they could impact on much wider learning outcomes. It was also clear that while certain pupils considered 'hard to reach' benefited from these different approaches, all young people improved their learning outcomes through these approaches.

9.11 Conclusions

At the heart of a School Citizens Assemblies process is the belief that by working together through a collaborative and co-creative process of research, innovation and action we can make a difference in the world around us. This relies on a process of divergent and convergent thinking and activities. While divergent activities open and disrupt spaces and processes to encourage different, innovative and creative views, perspectives and ideas to emerge. In contrast convergent activities, narrows down our focus and directs our thinking and practices to achieve certain actions, visions and outcomes in specific, systematic and well-ordered ways.

During this process of analysis and challenge-led learning, it is also important to unpack the different relational assemblages and black boxing that can limit or restrict an inclusive and equitable process of action and interaction. This involves moving between divergent and convergent processes, and critically reflecting on how certain processes of translation and the dynamics of divergence and convergence can impact on individuals, organisations and society. This includes the different aspects and relations that reside beneath the surface and acknowledging the potential impact these may have on the process of emergence and related outcomes (the impact of exam, assessment, university entrance requirements and accountability practices on school curriculums and teaching and learning practice; the way certain matters of fact (about climate change actions) are taken for granted and not opened-up as 'matters of concern'). A key aspect of the SCA process is creating and opening spaces to unpack 'matters of fact' and view them as 'matters of concern'. The SCA process enables more divergent thinking and a greater empathy and acknowledgement of different perspectives and thinking, while also creating spaces for convergence and effective actions in practice.

To conclude, the SCA process seeks to act as a 'force for good' by encouraging and supporting schools to act as hubs, assembling communities, experts and pupils to focus on specific issues or problems. This is particularly important given the many different and complex issues that impact on people and nature, both locally and globally (poverty, pandemics and disease, climate change, biodiversity and sustainability). Understanding and solving these problems requires people, organisations and governments to work together in creative, informed, inclusive, sustainable, responsible and compassionate ways. We need to work collaboratively to fully appreciate the complexity, diversity and difference that underlie these challenges and the knowledge, skills, understanding that will help us to find effective and sustainable actions and solutions. We also need enrich the knowledge, skills, values and understanding of young people and empower them as changemakers in their local and global communities.

Two main impacts of SCAs include: Firstly, creating spaces where young people, schools, experts and stakeholders can work together on real-world issues and make a positive and effective difference to the world around them. Secondly, educational environments that provide learning that is engaging, inclusive and challenging for all and supports the achievement of each child's future potential. Although assessment and accountability can place competing pressures on time, capacity and outcomes, we need to work together to find ways of achieving the best outcomes for all schools, teachers, pupils and society. Therefore, rather than trying to solve problems alone and in ways that lack compassion, critical thinking and creative problem solving, the SCA approach provides a way to research problems, generate ideas and tackle solutions in collective and collaborative ways. Learning how to work together to tackle complex problems facing us now and in the future is a crucial issue for individuals, organisations and society and we need approaches and practices within our educational system to support this through the range of effective approaches to teaching, learning and community engagement.

Bibliography

Akrich, M. (1992). *The de-scription of technical objects Shaping technology, building society: Studies in sociotechnical change* (pp. 205–224). MIT Press.

Aroles, J., & McLean, C. (2016). Rethinking stability and change in the study of organizational routines: Difference and repetition in a newspaper-printing factory. *Organization Science, 27*(3), 535–550.

Ball, S. (2008). The legacy of ERA, privatization and policy ratchet. *Educational Management Administration & Leadership, 36*(2), 185–199.

Barab, S. A., & Luehmann, A. L. (2003). Building sustainable science curriculum: Acknowledging and accommodating local adaptation. *Science Education, 87*(4), 454–467.

Cambridge Primary Review. (2009a). *Towards a new primary curriculum, a report from the Cambridge Primary Review, Part 1: Past and present*. Faculty of Education.

Cambridge Primary Review. (2009b). *Towards a new primary curriculum, a report from the Cambridge Primary Review, Part 2: The future*. Faculty of Education.

Cooren, F. (2018). Acting for, with, and through: A relational perspective on agency in MSF's organizing. In B. H. J. M. Brummans (Ed.), *The agency of organizing* (pp. 142–169). Routledge.

Cooren, F., Fox, S., Robichaud, D., & Talih, N. (2005). Arguments for a plurified view of the social world: Spacing and timing as hybrid achievements. *Time & Society, 14*(2–3), 265–282.

Central Advisory Council for Education (England). (1967). *Children and their primary schools (Plowden Report)*. HMSO.

Dewey, J. (1916). *Democracy and education: An introduction to the philosophy of education*. Macmillan.

Fenwick, T. (2011). Reading educational reform with actor network theory: Fluid spaces, otherings, and ambivalences. *Educational Philosophy and Theory, 43*(S1), 114–134.

Fenwick, T., & Edwards, R. (2010). *Actor-Network theory in education*. Routledge.

Gherardi, S., & Nicolini, D. (2005). Actor-Networks: Ecology and entrepreneurs, In Barbara Czarniawska and Tor Hernes (Eds.), *Actor-Network theory and organizing*. Liber

Haberman, M. (1991). The pedagogy of poverty versus good teaching. *Phi Delta Kappan, 73*, 290–294.

Hilliard, I. (2013). Responsible management, incentive systems, and productivity. *Journal of Business Ethics, 118*(2), 365–377.

Kidd, D. (2020). *A Curriculum of Hope: As rich in Humanity as in Knowledge*. Independent Thinking Press.

Kidd, D., & Roberts, H. (2018). *Uncharted territories: Adventures in learning*. Independent Thinking Press.

Knight, J. (2021). Education (Environment and Sustainable Citizenship) Bill. Lord Jim Knight. https://bills.parliament.uk/publications/41632/documents/306

Koh, J., Chai, C. S., Wong, B., & Hong, H. Y. (2015). Design thinking and education. In *Design thinking for education*. Springer. https://doi.org/10.1007/978-981-287-444-3_1

Laasch, O., & Gherardi, S. (2019). *Delineating and reconnecting responsible management, learning, and education: A research agenda through a social practices lens*. Paper presented at the Academy of Management Annual Conference, Boston.

Latour, B. (2011). "What's the story?" Organizing as a mode of existence. In J. H. Passoth, B. Peuker, & M. Schillmeier (Eds.), *Agency without actors? New approaches to collective action*. Routledge.

Latour, B. (2005). *Reassembling the social: An introduction to actor-network-theory*. Oxford University Press.

Latour, B. (2013). *An inquiry into modes of existence*. Harvard University Press.

Law, J. (1992). Notes on theory of the actor-network: Ordering, strategy, and heterogeneity. *Systems Practice, 5*(4), 379–393.

Leat, D. (2107). *Enquiry and project based learning*. Routledge.

Leat, D., Thomas, U., & Reid, A. (2012). The epistemological fog in realising learning to learn in European curriculum policies. *European Educational Research Journal, 11*(3), 400–412.

Lor, R. (2017). *Design thinking in education: A critical review of literature*. Asian Conference on Education & Psychology, Bangkok, Thailand, 24–26 May.

MacPherson, A., Kofinas, A. K., Jones, O., & Thorpe, R. (2010). Making sense of mediated learning, *Management Learning, 41*(3), July 2010.

Marble, S. (2012). Becoming-teacher: Encounters with the other in teacher education. *Discourse: Studies in the Cultural Politics of Education, 33*(1), 21–31.

Mason, C., Cremin, H., Warwick, P., & Harrison, T. (2010). *Building voice, civic action, and learning: Variation in levels of civic engagement amongst young people living in areas of socio-economic disadvantage*. A report of the Engaged project. Faculty of Education, University of Cambridge.

McAlister, S. (2013). Why community engagement matters in school turnaround. *Voices in Urban Education, 36*, 35–42.

McLean, C., & Aroles, J. (2016). Critical realism and actor-network theory/deleuzian thinking: A critical comparison in the area of information systems, technology and organizational studies. In L. Introna, D. Kavanagh, S. Kelly, W. Orlikowski, & S. Scott (Eds.), *Beyond interpretivism? New encounters with technology and organization. IFIP Advances in information and communication technology*. Springer.

Melles, G., Anderson, N., Barrett, T., & Thompson-Whiteside, S. (2015). Problem finding through design thinking in education. *Inquiry-based learning for multidisciplinary programs: A conceptual and practical resource for educators* (Innovations in Higher Education Teaching and Learning, Vol. 3), (pp. 191–209). Emerald Group Publishing Limited, Bingley. https://doi.org/10.1108/S2055-364120150000003027

Moberg, E. (2018). Exploring the relational efforts making up a curriculum concept—An actor-network theory analysis of the curriculum concept of children's interests. *Journal of Curriculum Studies, 50*(1), 113–125.

Nichols, M., & Cator, K. (2008). *Challenge based learning: Take action and make a difference. Challenge based learning white paper*. Apple.

OECD. (2018). *The future of education and skills: Education 2030*. http://www.oecd.org/education/2030/learning-framework-2030.htm

Painter-Morland, M. (2015). Philosophical assumptions undermining responsible management education. *Journal of Management Development, 34*(1), 61–75.

Rasche, A., & Gilbert, D. U. (2015). Decoupling responsible management education: Why business schools may not walk their talk. *Journal of Management Inquiry, 24*(3), 239–252.

Rose, J. (2009). *Independent review of the primary curriculum: Final report*. Nottingham, DCSF.

Singleton, D. V., & Law, J. (2005). Object lessons. *Organization, 12*(3), 331–355.

Tillbury, D. (2011). *Education for sustainable development: An expert review of processes and learning*. UNESCO.

Tsoukas, H., & Chia, R. (2002). On organizational becoming: Rethinking organizational change. *Organization Science, 13*(5), 567–582.

United Nations. (2016). *The sustainable development goals report*. United Nations.

United Nations Children's Fund (2019). UNICEF Education Strategy 2019–2030. Sept 2019. New York.

Venturini, T. (2010). Diving into Magma: How to explore controversies with actor-network theory. *Public Understanding of Science., 19*(3), 258–273.

Wals, A. E. J. (2012). *Shaping the education of tomorrow: 2012 Full-Length report on the UN decade of education for sutainable development*. UNESCO.

Chapter 10
Improving Corporate Governance Practices in the Catholic Church's Dioceses

José Lázaro Oliveira Nunes, Marcia Juliana d'Angelo, and Raysa Geaquinto Rocha

Abstract **Research Question/Issue:** This research has two objectives: (i) to discuss the level of adherence to corporate governance practices in Latin rite dioceses and (ii) to propose an ecclesiastical model of corporate governance.

Research Findings/Insights: We analyzed corporate governance practices in 98 Roman Catholic apostolic church dioceses in Brazil, Spain, and Italy. Its corporate governance model is based on the Code of Canon Law from 1983—a set of rules governing its acting. However, there are no relevant impediments to using well-consolidated governance practices in for-profit organizations, which can be 'canonized'. In particular, the principles of transparency (disclosure), equity (fairness), and accountability.

Theoretical/Academic Implications: The article develops a coherent, objective, and transparent statistical model, which can be applied on a worldwide scale, capable of seeking evidence of corporate governance in complex organizations, in terms of singularities—contingency (time and place) or paradox (conservative and progressive), as is the case with organizations with religious women.

Practitioner/Policy Implications: Our practical contribution is the new ecclesiastical corporate governance proposal—the Diocesan Corporate Governance Index (DCGI). It adds a fourth dimension—disclosure—to the traditional ecclesiastical governance model based on the Code of Canon Law, composed of three dimensions—conformity, mechanism, and performance. Adding disclosure dimension to the diocese's corporate governance model places decisions on how and when to put such a model into practice and excludes decisions on its feasibility and implementation.

Keywords Dioceses Corporate Governance Index · Corporate governance · Catholic Church · Latin dioceses · Disclosure

J. L. O. Nunes · M. J. d'Angelo
Fucape Business School, Vitória, Espírito Santo, Brazil

R. G. Rocha (✉)
NECE Research Center, University of Beira Interior, Covilhã, Portugal

D. Crowther and S. Seifi (eds.), *Preparing for a Sustainable Future*,
Approaches to Global Sustainability, Markets, and Governance,
https://doi.org/10.1007/978-981-99-2456-1_10

10.1 Introduction

Religion plays a relevant role in education, economics, and politics, showing no emptying of religiosity but a continuous permanence (Inauen et al., 2010). Thus, studies on religious organizations can generate information about a neglected area of 'non-commercial' behavior, which affects the attitudes and economic activities of individuals, groups, and even society.

In this sense, the Holy See—Vatican—recommends that the Catholic Church contribute to the concreteness of the 2030 Agenda for Sustainable Development, proposed by the United Nations (UN) organizations. It has 169 goals grouped into 17 objectives, to provide a more prosperous life for people and care for the planet (Vatican News, 2019a). For George, Howard-Greenville, Joshi, and Tihanyi (2016, p. 1880), this agenda covers part of the world's grand challenges: 'formulations of global problems that can be plausibly addressed through coordinated and collaborative effort'.

In particular, the Catholic Church can contribute to this agenda in four dimensions: (i) social field, focused on man's needs, through human promotion initiatives on all continents, (ii) in the environmental area, aimed at the conservation and protection of the environment; and (iii) in economics, in addressing the sustainable use of natural resources. Pope Benedict XVI, in 2008, added social inequality and environmental pollution among the new capital sins, for which Christians must apologize. Thus, the fourth and last dimension of the Catholic Church's contribution is institutional, which can put this agenda into practice through initiatives with an ethical and spiritual focus to develop the human being who runs organizations (Vatican News, 2019a).

However, like for-profit organizations that face corruption, accounting, and ethical scandals, religious organizations also face problems and challenges related to governance. Among them, bankruptcy filings in Catholic dioceses (Downes, 2015); sex scandals (Barth, 2010); concern with balancing sacred and secular performance and goals (Payer-Langthaler & Hiebl, 2013); members' dishonesty (Rost et al., 2010); embezzlement of own or government funds for other purposes; and the concern with fiscal supervision and financial management (Elson et al., 2007). For example, in Brazil, there are recent accusations of misuse of funds by the Rector of the Sanctuary Basilica of the Divine Father Eternal, Father Robson Pereira (Oliveira, 2020), and in Italy, the money laundering complaints at the Institute for Works of Religion (IOR), the Vatican Bank (Verdú, 2019). On the other hand, in Spain, there is a corruption scandal in the Spanish parish of Borja (Europa Press, 2013).

Adding to the international financial crisis that started in 2008, these scandals reveal a crisis of credibility and, consequently, a crisis of corporate governance in organizations (Inauen et al., 2010). The absence of 'processes, customs, policies, laws, and institutions affect how an organization is directed and controlled' (Inauen et al., 2015, p. 764). Corporate governance is a system that involves regulation, control bodies, and several social actors (society in general, partners, directors, employees, suppliers, fiscal and administration councils, family councils, among others). The focus is on generating value in the long run (IBGC, 2014). In other words, 'the system by which companies are directed and controlled. Boards of directors are responsible for the governance of their companies' (Tihanyi et al., 2015, p. 1).

The Catholic Church has its corporate governance model, based on the Code of Canon Law—a set of rules that govern life in this organization (John Paul II, 1983). This model is well defined and anchored in three variables: (i) mechanism—diocesan curia, councils of priests, college of consultants; (ii) conformity—the bishop's three-fold mandate concerning Jesus, which is to teach, sanctify and govern; and (iii) performance—bishops' pastoral ministry seen from the number of sacraments (Pfang, 2015). However, there is no mention of the practice of disclosure, the voluntary provision of information to its external and internal stakeholders.

Despite its rigid rules and laws, the Church is an open system, which allows the use of sound governance principles already successfully employed in for-profit organizations, which can be 'canonized' like past actions: those of St. Augustine, with Plato's 'pagan' philosophy, and of St. Thomas Aquinas, with Aristotle's 'pagan' philosophy. Thus, this study argues an opening in the Catholic Church for the governance practices of disclosure (transparency), fairness (equity), and accountability. If the adoption of mechanism, compliance, and performance practices already contributes to balancing its spiritual mission and making its management effective (Pfang, 2015), adopting disclosure practices contributes to strengthening its management. The Catholic Church's crises indicate the need for changes in its structure, organizational processes, and governance to effectively manage eventual crises (Barth, 2010). For example, in 2019, Pope Francis renewed the Institute for the Works of Religion (*Istituto per le Opere di Religione* or IOR), commonly known as the Vatican Bank, statutes so that external auditors—individual or legal entity—can check its balance sheets (Vatican News, 2019b).

Given the above, this research has two objectives: (i) to discuss the level of adherence to corporate governance practices in Latin rite dioceses—Brazil, Spain, and Italy and (ii) to propose an ecclesiastical model of corporate governance.

Although the Catholic Church's legal personification is the Vatican (Holy See) and the Pope, its supreme leader, aided by the Roman Curia with its congregations, secretariats, courts, and other bodies, the administration of the Catholic Church takes place in dioceses. The bishop makes most administrative decisions; he is the dioceses' 'president', and the 'CEO' (Chief Executive Officer). For example, establishes the Code of Canon Law (CIC) can. 381, § 1: 'The diocesan Bishop, in the diocese entrusted to him, has all the ordinary power, proper and immediate' (Barth, 2010; Cavalcante, 2018; João Paulo II, 1983; Pfang, 2015).

There are 2,998 dioceses in the world, resulting from the territorial division established by the Pope. Each diocese—also called the Particular Church—delimits the bishop's area of activity and government. In particular, in Latin rite dioceses, such as Brazil, 277 dioceses make up the National Conference of Bishops of Brazil (CNBB), with 172.2 million faithful. In addition, there are 70 dioceses in Spain, forming the Spanish Episcopal Conference (CEE), with 43.3 million Catholics. Finally, in Italy, there are 226 dioceses organized at the Italian Episcopal Conference (CEI), with 58 million faithful (The Vatican, 2018).

Despite common sense, dioceses work to spread the faith and not deal only on the spiritual plane. For example, the National Conference of Bishops of Brazil (CNBB), through Caritas International, is a network and has more than 15 thousand agents among employees and volunteers, dealing with considerable human and financial resources. Over the past ten years, it has assisted more than 300,000 families (CNBB, 2019). The Spanish Episcopal Conference (CEE) announced, in 2016, that the dioceses contributed around 70 million euros to 373 conservation and rehabilitation projects and that 3,168 properties of cultural interest belong to the Church (CEE, 2019). The Italian Episcopal Conference (CEI) informed that 95 projects for Asia, Africa, and Latin America, in education, health, and social rehabilitation, were among its priorities, adding up to a total of 16 million euros (CEI, 2019).

Studying this reality can provide new insights, besides the religious aspects of faith, spirituality, or morals. It encompasses organizations' different economic and strategic aspects from commercial or governmental ones. Address religious organizations only concerning the spiritual field is to ignore history. For example, such organizations were instrumental in cities' economic systems developed around monasteries, cathedrals, or shrines in the West (Inauen et al., 2010). Pfang (2015, p. 43) stated, 'good governance practices are applicable and relevant to the Catholic Church'.

Previous studies (Bazanini & Machado Junior, 2018; Downes, 2015; Elson et al., 2007; Faller & Zu Knyphausen-Aufseß, 2016; Giumbelli, 2011; Inauen et al., 2010; Murtinho et al., 2018; Pfang, 2015; Pless et al., 2017) addressed many aspects of religious organizations—inside and outside the Christian world, inside and beyond the Catholic Church—in its various aspects. However, gaps such as quantitative research in dioceses and corporate governance practices remain because it is an extensive and complex subject due to its organizational scope, nature, and idiosyncrasies. Furthermore, unlike Monasteries (Benedictines, Carmelites, Poor Clares) or Religious Orders (Franciscans, Jesuits, Dominicans), whose archives usually are well structured, with more comprehensive availability of people and accessibility, dioceses lack this level of organization, which favors research on this topic (Inauen et al., 2010).

Given the above, the study aims to fill these gaps in quantitative studies and present new reflections of this ecclesial and 'non-commercial' reality regarding the Apostolic Catholic Church Rome's dioceses' governance. Furthermore, George et al. (2016) also recommend comparative studies on governance that consider global dimensions and social impacts. Hence the proposal for comparing corporate governance practices between the dioceses in Latin rite—Brazil, Spain, and Italy.

The study's main contribution is to propose the Diocesan Corporate Governance Index (DCGI), focusing on four dimensions—disclosure, compliance, management mechanisms, and performance. It is a new ecclesiastical model of corporate governance for dioceses, considering the practical implications, technological innovations, and recommendations of well-consolidated practices, for example: 'do or explain'. And in the face of research evidence, the need for change and adaptation in ecclesial management. Thus, based on mechanism, compliance, and performance dimensions, the existing model is expanded by inserting the disclosure dimension: transparency and accountability. The proposed model seeks to minimize imperfections by focusing on corporate governance in the Catholic Church's dioceses, not restricting the fulfillment of the diocesan Bishop's spiritual mission, responsible for governance, but contributing to the greater effectiveness of governance.

This proposal advances the traditional model based on the 1983 Code of Canon Law, based only on three dimensions: compliance, management mechanisms, and performance (Pfang, 2015). The dioceses' traditional corporate governance model does not correspond to the current needs generated by the credibility crises due to corruption scandals, fiscal supervision, and financial management. Therefore, it demands improving the corporate governance model, emphasizing the disclosure dimension (accountability and transparency).

10.2 Theoretical Background

10.2.1 Corporate Governance in Different Sectors

In economic and social agents, the first sector is the State, which is public; the second is the market, which is private and aims at a profit. However, with the worsening of social inequalities and the State's inability to effectively promote social justice, a third sector has emerged, encompassing many organizations such as associations, foundations, unions, cooperatives, churches, and non-governmental organizations (NGOs). These are private, non-profit initiatives that offer public goods and services in the social (including health and education), ecological, and spiritual (Ávila & Bertero, 2016; Borzaga & Fazzi, 2014; Kroeger & Weber, 2014; Schmidt, 2014; Pless et al., 2017; Windrum, 2014).

The adoption of Corporate Governance practices had high importance in for-profit organizations, especially after the scandals arising from corporate fraud, which caused financial loss to shareholders, the bankruptcy of companies, destruction of jobs, and the opening of a large number of criminal cases in several countries (Almeida & Zagaris, 2015; Ávila & Bertero, 2016; George et al., 2016; Millar et al., 2005; Rezende & Dalmácio, 2015). That is why the logic and influence of the market mark studies on corporate governance. So much so that commercial language stands out in its most common definitions, according to some market organizations—the Brazilian Institute of Corporate Governance (IBGC, 2015), in Brazil; the National Securities Market Commission (CNMV, 2016), in Spain; the Internal Controls System (SCI), in Italy (Mariani et al., 2013). Thus, for-profit companies follow the principles and guidelines of transparency (disclosure), equity (fairness), accountability, compliance, and social responsibility.

However, corporate governance is no longer exclusive to private companies but is present in governmental and non-governmental organizations (CNMV, 2016; Mariani et al., 2013). For Barraket and Yousefpour (2013, p. 447), the practice of 'communicating organizational impacts to a variety of stakeholders has gained increasing importance within all sectors'. For the Brazilian Institute of Corporate Governance

(IBGC), corporate governance comprises public bodies and non-profit organizations with no owners. Hence the term institutional governance is used as a synonym for Corporate Governance to encompass non-profit organizations (IBGC, 2014).

Public administration follows the precepts of governance in private organizations (Kissler & Heidemann, 2006). It is monitored and charged for providing more efficient and effective services to citizens, that is, performance (Chiu, 2016; Yong & Wenhao, 2012). It adopts practices of transparency (Ladeur, 2017), horizontal accountability by control bodies (Wolff, 2018), and societal accountability by citizens and third sector organizations (Faller & Zu Knyphausen-Aufseß, 2016; Heinrich & Brown, 2017).

Some studies have shown the pressure for third-sector organizations to be more accountable for the final results (accountability) (Acar et al., 2012; Greiling & Stötzer, 2016; Grizzle & Sloan, 2016). Stakeholders are also charging these organizations for higher quality, efficiency, and effectiveness in providing social services (Benjamim & Campbell, 2015; Melão et al., 2017). As they are dependent on public and private donations, they deal with several stakeholders (Irene et al., 2016).

In particular, adherence to Corporate Governance in the Catholic Church, shaken by scandals and lack of transparency (Barth, 2010; Downes, 2015; Rost et al., 2010), can contribute to regaining stakeholder reliability and provide an organizational advantage strategically for the continuity and efficiency of its mission, such as for-profit organizations. Therefore, by adhering to well-consolidated corporate governance practices, religious organizations can assure the faithful of the correct and effective use of the donated resources, curb corrupt practices, making the institutional image solid and reliable (Inauen et al., 2010; Rezende & Dalmácio, 2015).

One cannot ignore the risks and difficulties when dealing with complex issues such as corporate governance in religious organizations, which are not used to such models of control, monitoring, and language. Remember, such knowledge was structured and thought from commercial and profit logic (Lugoboni et al., 2018).

Thus, in the face of the mistrust and lack of credibility of many non-profit organizations, particularly faith-based organizations, corporate governance becomes a compelling and legitimate strategic differential that provides credibility and transparency. Adhering to the concepts of corporate governance is no longer a simple choice for religious organizations and becomes fundamental for the organizational future (Ávila & Bertero, 2016).

10.2.2 Administrative Structure and Governance Practices in the Catholic Church

The Catholic Church is one of the largest and oldest organizations globally, with more than twenty centuries. An organization with such a long time of effectiveness undergoes considerable changes in its administration. Although it considers itself absolute in many things, it treats its administration in a very relative and plural way. The Catholic Church has preserved many secular rites and functions, but its administration has proved contingent and even paradoxical over the centuries since each diocese has administrative freedom to conduct its management (Cavalcante, 2018; Elson et al., 2007; Giumbelli, 2011; Messi, 2010).

The Catholic Church's administrative structure has three vertical levels linked to an agent (Pope, Bishop, and priest) and a geographical territory (world, diocese, and parish). The first level is the Holy See, entrusted to the Pope, the Supreme Pontiff. Bishops' Conferences act as a collegiate of equals. In Brazil, it is the National Conference of Bishops of Brazil (CNBB); in Italy, it is the Italian Episcopal Conference (CEI); and in Spain, it is the Spanish Episcopal Conference (EEC). In countries where the Holy Church has diplomatic relations, Apostolic Nunciatures are equivalent to an embassy. In Dioceses and Parishes, there are several types of Councils: that of Economic Affairs, that of Priests, that of Consultants, that of Pastorals and others, that assist in the government of the Church (Cavalcante, 2018; Messi, 2010; Pfang, 2015; Vatican, 1962).

The second level includes the Private Churches (archdioceses, dioceses, prelates, abbeys, and others similar to dioceses) entrusted to the Bishop' pastoral/administration, chosen from among the priests and appointed by the Pope. Anyone over 35 years old and who has been a priest for more than five years can be appointed a Bishop. Upon completing 75 years old, the Bishop must present his resignation to the Pope (Cavalcante, 2018; João Paulo II, 1983, p. 71). Bishops are not deputies of the Pope, as each has the same authority as the Pope in his diocese concerning the entire Catholic Church. It is the responsibility of the Diocesan Bishop to govern the church entrusted to him, with legislative, executive, and judiciary power. The Pope is the successor of Peter and the Bishops of the Apostles. That is why the diocesan Bishop is involved from a legal perspective whenever there are scandals in his diocese (Cavalcante, 2018; João Paulo II, 1983, p. 58; Pfang, 2015).

The third level covers parishes, a subdivision of the diocese. Can. 515, § 1, of Canon Law sets the parish that is stably established in the Particular Church. The parish priest undertakes pastoral care under the authority of the diocesan Bishop (John Paul II, 1983, p. 95).

When prescribing the Catholic Church's organizational structure, the Code of Canon Law presents its institutional governance, organized in three dimensions. They are mechanisms, compliance, and performance (Pfang, 2015). By mechanisms (canon 469 to canon 514), it is understood that the diocesan Bishop has a body to assist him in the administration of his diocese. This management body is the 'Diocesan Curia'. It comprises people and institutions that assist the bishop in pastoral, spiritual, disciplinary care in the communicative, administrative, judicial, and other fields. The mandatory agents that make up the diocesan curia mechanisms are the chancellor, financial council, vicar general, Presbyterian council, and consultants' college. The optional agents are episcopal vicar, episcopal council, and moderator).

The dimension compliance (Can 386, 387, and 391) means the mission of the diocesan Bishop in his relationship of conformity with the threefold offices of Christ: prophet-teaching, priest-sanctifying, and king-governing. The Bishop is thus personally responsible for this aspect of the Church's mission in his diocese. In other words, each Bishop's mission is to teach (can. 780), through homilies and catechesis; sanctify (can. 387), through the liturgy and the sacraments; and govern (can. 391), exercising legislative, executive, and judicial power (Cavalcante, 2018; João Paulo II, 1983; Messi, 2010; Pfang, 2015).

The dimension performance shows the Catholic Church's specific mission is spiritual nature, namely: the salvation of souls, according to can. 771 and 761 (John Paul II, 1983). However, how to translate this performance into numbers: would it be for the salvation of souls? The Church uses statistics, accounting for and describing the sacraments administered: baptisms, confessions, Eucharists. Masses, confirmations, marriages, anointing the sick, and sacraments of order (Cavalcante, 2018; Messi, 2010; Pfang, 2015).

The Catholic Church's institutional governance uses to monitor the three dimensions—mechanism, compliance, and performance—mainly through the canonical visit and the ad *limina apostolorum* visit. The canonical visit, also called the pastoral visit, follows the norms by canons 396 to 398, which impose the Bishops a mandatory visit in each parish annually. Within five years, the entire diocese receives a pastoral visit. Among the visit's purposes, the administrative and control function: activities planning and finance. The canonic visit is analogous to a business audit (Cavalcante, 2018; João Paulo II, 1983).

The *ad limina apostolorum* visit, which means on the threshold, on the limits (of the basilicas) of the apostles (Peter and Paul), is a visit by the bishops 'to the tombs of the Apostles', that is, to the Pope, which must take place every five years. On that occasion, the Bishop has to present the Quinquennial Report, which covers the diocese's entire life, to inform Pope about each particular church's management, as provided in the can. 399 and can. 400 (Battisti, 2010; Cavalcante, 2018; João Paulo II, 1983). There are three central moments of the visit: the pilgrimage to the tombs of the two apostles (Pedro and Paulo); the personal encounter with the Roman Pontiff; and the colloquiums, individually or collectively, with the departments of the Roman Curia (Cavalcante, 2018; Gomes, 2009).

Both Gomes (2009) and Battisti (2010) recall that six months earlier (not less than three), the Bishop must send the Five-Year Report to the Holy See. Unfortunately, these reports are not published, as they are the Bishops' documents. However, why not disclose performance data to provide useful information for pastoral planning and dioceses' stakeholders? Despite this restriction, it is possible to access the *Formula Relationis Quinquenalis*, published by the Congregation of Bishops. This document gives clear instructions on the structure and information required in the report. It consists of 23 chapters, each evaluating an aspect of the diocese and pastoral performance in the period.

10.3 Methodology

10.3.1 The Proposed Diocesan Corporate Governance Index (DCGI)

Coy and Dixon (2004) proposed the Public Accountability Index (PAI) model, a dissemination index with parametric statistical properties and polychotomic approaches developed to overcome standards with limited (dichotomous) approaches. The new model became the basis for the development and formulation of several corporate governance indexes in Brazil and abroad, such as the Corporate Governance Index (CGI), by Silveira et al. (2008) and Silveira (2005); Corporate Governance Index (CGI), by Silva and Leal (2005); Brazilian Corporate Governance Index (BCGI), by Lopes and Walker (2008); and the Corporate Governance Index for Football Clubs (CGIFC), by Rezende et al. (2010).

The Diocesan Corporate Governance Index (DCGI), proposed in this study and shown in Fig. 10.1, is based on the Public Accountability Index (PAI) by Coy and Dixon (2004), and on the Corporate Governance Index in Football Clubs (CGIFC), by Rezende et al. (2010) and Rezende and Dalmácio (2015). It covers four dimensions of corporate governance: disclosure, compliance, mechanism, and performance. The composition of the DGGI Index is the result of the analysis of adherence to these dimensions by the dioceses (IBGC, 2014; João Paulo II, 1983; Pfang, 2015; Rezende & Dalmácio, 2015; Rezende et al., 2010).

10.3.2 Survey

The Diocesan Corporate Governance Index (DCGI) operationalization uses primary data from a questionnaire with 40 questions. The disclosure construct (transparency and accountability) has 13 items from the Code of Best Practices of the Brazilian Institute of Corporate Governance (IBGC), the resolutions of the Federal Accounting Council (CFC,), and the studies by Rezende et al. (2010) and Silva and Leal (2005). The compliance construct has nine items also from the Code of Best Practices of the Brazilian Institute of Corporate Governance (IBGC) and the studies by Oliveira et al. (2013) and Silva and Leal (2005). The mechanism construct (effectiveness) has eight items based on the Apostolic Constitution '*Sacrae Disciplinae Leges*', by João Paulo II (1983) and the studies by Oliveira et al. (2013), Rezende et al. (2010), Silva and Leal (2005). Finally, the performance construct (organizational continuity) has ten items based on the Apostolic Constitution '*Sacrae Disciplinae Leges*', by João Paulo II (1983), the Code of Best Practices of the Brazilian Institute of Corporate Governance (IBGC), and the study by Rezende et al. (2010). For each question, respondents had to choose answers between yes and no.

Since Coy and Dixon (2004) created the Public Accountability Index (PAI), there is an awareness that an index's construction is full of subjectivity. Thus, as dioceses

Diocesan Corporate Governance Index (DCGI)			
	Corporate governance practices		Evaluation measures
Disclosure	The principle of transparency and accountability governs it. The former comprises the free and spontaneous disclosure of information to interested parties inside and outside the organization. The latter recalls a moral and even a contractual obligation to render accounts of the acts performed during the exercise of their mandates. There is an obligation to publish accounting reports required by law, such as the annual report. Corporate governance to be effective depends to a large extent on the active participation of those responsible for the organization (Coy and Dixon 2004; Elshandidy and Neri 2015; Rezende and Dalmácio 2015; Rezende et al. 2010; Soltani and Maupetit 2015).	Transparency, Accountability. Accountability for results. Information disclosure to the Catholic Church's internal and external stakeholders out of moral, not legal, obligation.	Through periodic reports, use international accounting standards, independent external auditing, operating and independent counsels.
Mechanism	It is ruled by the principle focused on effectiveness, people, and institutions at the institutional objective's service. First, starting from its competence to produce the best yield with minimal errors and little expenditure (efficiency). Second, fulfilling the planned goals and objectives, that is, being able to achieve the expected results (effectiveness) (Silva and Leal 2005; João Paulo II 1983; Kobuthi, K'Obonyo, and Ogutu,2018; Manzaneque, Priego, and Merino 2016; Pfang 2015; Rezende and Dalmácio 2015).	Effectiveness of the diocesan curia. Produce the best performance with minimal errors and little expenditure (efficiency). It means compliance with planned goals and objectives (effectiveness). Permanently monitors and guarantees discipline, governance and administration practice in the local Catholic Church.	Through the activities and organs stipulated in canon 469, (John Paul II, 1983). Also include other organs, apparatus, methodologies, and offices that the Bishop deems necessary for his diocese government, provided that they are not parallel structures to those provided in the Code of Canon Law.
Conformity	Governed by the principle of harmonization (Prado, Souza, Raymundo, and Horizonte, 2019), that is, act according to the regulatory rules prescribed in the laws, such as bylaws, internal regulations, civil regulatory institutions of the State. It is a system that cuts across all levels of the organization. It is a proposal for a holistic view that encompasses organizational identity, governance agents, and other compliance elements. The organization's identity determines ethical deliberation and guides the actions of agents and the functioning of the compliance system, which seeks to meet three primary objectives: prevent, detect and respond (Bevir 2006; Elshandidy and Neri 2015; Enriques and Volpin 2007; Hussain, Rigoni, and Orij 2018; Brazilian Institute of Corporate Governance (IBGC) 2017; John, De Masi, and Paci 2016; Soltani and Maupetit 2015).	Acting in the fulfillment of Christ's threefold office: teaching, sanctifying and governing, according to civil and canonical regulatory norms. It is a system that cuts across all diocese levels, a proposal for a holistic view.	The external audit will analyze the application and harmony of canonical and civil norms in the diocese organization and practices. In addition to seeking evidence of the primary compliance objectives: prevent, detect and respond.
Performance	It is ruled by the principle focused on organizational continuity, on its perpetuity. It takes place through a long-term vision regarding the sustainability of human and environmental resources, as well as in the clear definition of the business and organizational operations vision, avoiding putting organizational continuity at risk (Aráujo, Mendes, and Lustosa 2012; Brazilian Institute of Corporate Governance (IBGC) 2017; João Paulo II 1983; Pfang 2015; Rezende and Dalmácio 2015).	Organizational continuity, its perpetuity, occurs through long-term planning for the sustainability of human, environmental and financial resources. Clear definition of the mission and vision of pastoral plans, formation, initiation and belonging defined in the Diocesan Assemblies / Synods.	Through statistical data and analysis of satisfaction of the sacraments, more precisely two: Baptism, by which one begins to be part of the Church; Order, one that guarantees the continuity of the mission through the ministers (deacon, priest and Bishop).

Fig. 10.1 Description of the Diocesan Corporate Governance Index (DCGI) (*Source* The Authors based on Pfang [2015])

are not-for-profit organizations, it was necessary to adapt the questions to the context of each country and dioceses to mitigate the subjectivity of respondents (Rezende & Dalmácio, 2015; Rezende et al., 2010) and to be able to build the Diocesan Corporate Governance Index (DGCI).

A pre-test was conducted with 26 respondents from Brazilian dioceses to assess the questions' relevance, meaning, and clarity. Based on the responses and the evaluation, the instrument was adjusted accordingly. The questionnaire was then translated from Portuguese into Spanish and Italian, seeking the full equivalence between languages due to the differences in the three countries' social, political, cultural, and legal contexts. Nevertheless, the Latin rite is based on the same canonical norms. See the appendix for more details.

After the pre-test stage, the questionnaire was made available to Brazil, Spain, and Italy's dioceses through the Google Forms application from June to September 2019. In addition, it was sent via email and WhatsApp application. The authors also made phone calls to collect data. The average response was 18.6% across the three countries, and the respondents' positions comprise the Bishop, the general vicar, and the bursar (treasurer), the ones most involved in the dioceses' management. Table 10.1 details this information.

To determine the weight of the constructs (Table 10.2), the authors used the inverse result criterion, reaching the ratio between the total questions in one dimension and the questionnaire's total questions. To restrict the bias of dimensions that have a higher number of questions to measure their concept, the authors rounded the results so that all dimensions had the same score, because the higher the number of questions to measure a dimension, the lower the individual weight each question (Rezende et al., 2010). Therefore, the final score is the number of weights in the constructs' indicators—evidence, conformity, mechanism, and performance—without possible bias due to the different number of questions (Cenfetelli & Bassellier, 2009).

When determining the average weight of these constructs, the model is inspired by the logic of Beattie et al. (2004), which focuses on individual dimensions and

Table 10.1 Sample data—Brazil, Spain, and Italy

	Brazil	Spain	Italy
National Conference of Bishops	CNBB	CEE	CEI
Number of Catholic faithful (millions)	172, 2	43,3	58,0
Total of Dioceses	277	70	226
Answers received	54	18	26
Return rate (%)	19,5	25,7	11,5
Respondents (Bishop)	17	4	9
Respondents (Vicar General)	21	8	12
Respondents (Treasurer)	16	6	5

Note CNBB: National Conference of Bishops of Brazil; EEC: Spanish Episcopal Conference; CEI: Italian Episcopal Conference.
Source Pontifical Yearbook (2018) and *Annuarium Statisticum Ecclesiae* (2016)

Table 10.2 Indicators and determination of model weights

Constructs	Items	Formula	Sum of items	Weight
Disclosure	E1 to E13	$E=\frac{13}{40}=0,33 => \frac{1}{0,33}=$ $3,03 \times 13 = 39{,}39 \cong$ $\frac{40}{13}=3,08$	(3,08 × 13) = 40,04 ≅ 40	3,08
Compliance	C1 to C9	$C=\frac{9}{40}=0,23 =>$ $\frac{1}{0,23}=4,35x9=$ $39,15 \cong \frac{40}{9}=4,44$	(9 × 4,44) = 39,96 ≅ 40	4,44
Mechanism	M1 to M8	$M=$ $\frac{8}{40}=0,20 => \frac{1}{0,20}=$ $5,00x8=40 \cong \frac{40}{8}=5$	(8 × 5) = 40	5,00
Performance	D1 to D10	$D=\frac{10}{40}=0,25 =>$ $\frac{1}{0,25}=4,00x10 \cong \frac{40}{10}=4$	(10 × 4) = 40	4,00
Maximum score (4 × 40 = 160)				160

Note Items refer to the questionnaire questions.
The formula is $PD = \frac{N\_QD}{TQ\_IGCD}$. PD represents weight by dimension; N._QD represents the number of questions per dimension of the DCGI; TQ_IGD refers to the questionnaire's total number of questions (Rezende & Dalmácio, 2015; Rezende et al., 2010)

their interrelations and the rationale of the Herfindahl or HHI Index, used to measure competition in the market. The higher the HHI, the lower the spread in the market and vice versa. Therefore, the greater the number of items in the construct, the lower its average weight.

For each dimension, diocese, and country, the next step was the sum of all questions with answers 'yes'. For example, in Brazil, the DBr37 diocese answered yes to six questions in the Disclosure dimension. Therefore, the score was 18.48 (6 × 3.08 = 18.48). In the Compliance dimension, he answered yes to four questions. Therefore, the score was 17.76 (4 × 4.44 = 17.76). In the Mechanism dimension, he answered yes to six questions. Therefore, the score was 30.00 (6 × 5.00 = 30.00). Finally, in the Performance dimension, he answered yes to six questions. Therefore, the score was 24.00 (6 × 4.00 = 24.00).

Each diocese's total score was calculated by adding the dimensions (disclosure, compliance, mechanism and performance) scores. For example, in Brazil, the total score of the DBr37 diocese was 90.24 (18.48 + 17.76 + 30.00 + 24.00 = 90.24). From the total score, each diocese was classified in three countries - Brazil (54), Italy (26), and Spain (18). Thus, the final classification comprises 98 positions (54 + 26 + 18 = 98).

Finally, the Diocesan Corporate Governance Index (DGCI) 's central tendency measures—mean, mode, and median—were calculated for each dimension proposed in this study, including the overall score and country's total score. These measures locate the center of the expected sample distribution (Hair et al., 2003).

The mean refers to the arithmetic average calculated by dividing the sum of the scores of each diocese by the number of dioceses in each country (Hair et al., 2003). For example, in Brazil (56 dioceses), the sum of the total score of each diocese was 5400.48 (DBr1 = 150.76 + DBr2 = 144.60 + ... + DBr54 = 49.60). Therefore, the mean score was 100.01 (5400.48/56 = 100.01). In Italy (26 dioceses), the sum of the total score of each diocese was 2752.56 (DIt1 = 140.56 + DIt2 = 134.32 + ... + DIt26 = 66.32). Therefore, the mean score for Italian dioceses was 105.87 (2,752.56/26 = 105.87). In Spain (18 dioceses), the total score for each diocese was 2080.36 (SD1 = 128.44 + SD2 = 127.52 + ... + SD18 = 95.52). Therefore, the mean score for Spanish dioceses was 115.58 (2080.36/18 = 115.58).

The mode is the measure of central tendency that identifies the value that occurs most frequently in the sample distribution or the most repeated value in a range of data. The median is the value that is at the center of the distribution or data range. It is the value below (and above) from which half of the values fall. If there is an even number of observations, the median is the average of the two central values. If there is an odd number of observations, the median is the central value (Hair et al., 2003). Thus, the following analyzes are based on the median.

10.4 Data Analysis

As shown in Table 10.3, none of the dioceses surveyed reached the maximum score of 160 points on the Diocesan Corporate Governance Index (DCGI). In Brazil, the one with the best performance (DBr1) obtained 150.76 points, which represents 94% of adherence to the requirements of the DCGI Index, also ranking first in the overall ranking. In Italy, the diocese that stood out the most (DIt1) reached 140.56 points, fulfilling 88% of the DCGI Index requirements and being in third place overall. It should be noted that the second place is also from a Brazilian diocese. In Spain, the one with the best performance reached 128.44 points (DEs1), indicating an adherence of 80% of the DCGI Index and occupying the ninth overall position.

The general median indicates that 50% of dioceses reached 105.22 points or 66% adherence to the requirements of the DCGI Index. In Brazil and Italy, the medians of 100.04 points and 108.84 points indicate that the dioceses fulfill 63% and 68% of the DCGI Index, respectively, while the median of Spain, which is slightly higher

Table 10.3 Centrality measures of Diocesan Corporate Governance Index (DCGI)—overall score and country's total score

	Overall score	Country's score		
		Brazil	Spain	Italy
Mean	104,72	100,01	115,58	105,87
Mode	119,88	100,32	119,88	108,84
Median	105,22	100,04	118,08	108,84

(118, 08 points), points to a 74% compliance with the Index. This evidence shows that these dioceses still need to invest in more actions for more effective governance.

The lowest performances in Brazil (DBr54 = 49.50 points) and in Italy (DIt 26 = 66.32 points) indicate that these dioceses comply with 31%, and 41% of the DCGI Index, respectively, showing low adherence to corporate governance practices. Hence, they are still far from the dioceses that fulfill 66% of this Index. On the other hand, in Spain, the diocese with the lowest performance (SD18 = 95.52 points) complies with 60% of the DCGI Index. Thus, Spanish dioceses are more committed to governance.

In Table 10.4, the Disclosure dimension's general median indicates that 50% of the dioceses reached 21.56 points, fulfilling 54% of the requirements for accountability to interested parties. The Italian and Spanish dioceses perform better, occupying the first twelve positions, and the Italian diocese (DIt1 = 40.04 points) fulfilled 100% of the disclosure requirements. The others served 85%, with 33.88 points.

The best performing Brazilian dioceses occupy the 13th (DBr1), 14th (DBr5), and 15th (DBr6) positions, each with 30.80 points, equivalent to 77% of compliance with the requirements of this dimension. The lowest performances are in Brazil, with two dioceses (DBr 47 and DBr 49) with 9.24 points, fulfilling 23% of that dimension and

Table 10.4 Centrality measures of Diocesan Corporate Governance Index (DCGI) dimensions—overall score and country's total score

	Disclosure			
	Overall	Brazil	Italy	Spain
Mean	22,47	18,59	25,00	30,46
Mode	21,56	21,56	27,72	33,88
Median	21,56	18,48	27,72	30,80
	Compliance			
	Overall	Brazil	Italy	Spain
Mean	20,66	19,32	21,00	24,17
Mode	17,76	17,76	17,76	31,08
Median	17,76	17,76	17,76	26,64
	Mechanism			
	Overall	Brazil	Italy	Spain
Mean	31,63	32,04	30,38	32,22
Mode	30,00	30,00	30,00	30,00
Median	30,00	30,00	30,00	30,00
	Performance			
	Overall	Brazil	Italy	Spain
Mean	29,70	30,06	29,65	28,72
Mode	32,00	40,00	32,00	32,00
Median	30,00	29,00	31,00	28,00

four dioceses (DBr46, DBr 52, DBr 53 and DBr 54) with 6.16 points that is, meeting only 15% of the disclosure requirements. These results reveal a low adherence and the need for more actions that favor the dissemination of information to interested parties, internal and external, to the dioceses (46%) that are in the second half of the table of the disclosure dimension.

The Compliance dimension's median indicates that 50% of the dioceses reached 17.76 points, fulfilling 44% of the requirements, showing a low adherence to civil and canonical regulatory standards. Two from Brazil (DBr1 and DBr2) and one in Italy (DIt1) reached 39.96 points, 100% fulfilled with these standards. As a negative highlight, a diocese in Italy (DIt26) did not score. The Brazilian dioceses (DBr22, DBr43, DBr45, DBr51, and DBr54) have the lowest score (4.44 points), fulfilling only 11% of the compliance requirements. The Spanish diocese with the best performance has 31.08 points, or 78% fulfilled with that dimension, while the one with the lowest performance has 8.88 points or 22% fulfilled with conformity requirements.

The Mechanism dimension's median shows that 50% of the dioceses reached 30 points, fulfilling 75% of these requirements, which refer to the services provided by the diocesan curia to the people and institutions in the service of the diocesan Bishop, revealing a high adhesion of the dioceses. Important to highlight the lowest performance is of five Brazilian dioceses (DBr9, DBr45, DBr50, and DBr54) and one Italian (DIt20), with 25 points, or 63% of meeting the requirements of this dimension, and the DBR54 diocese scored 20 points or fulfilled 50% of the provision of services.

In this dimension, the performance of Spanish dioceses is also noteworthy—ten organizations (DEs6, DEs8, DEs9, DEs10, DEs11, DEs13, DEs14, DEs15, DEs16, and DEs17) are in the last positions (from 84th to 93rd positions) with 30 points or 75% of compliance with the mechanism requirements. The highlights are eight Brazilian dioceses (DBr1, DBr2, DBr3, DBr4, DBr6, DBr8, DBr10, and DBr49) and one Italian (DIt2), which reached 40 points, or 100% adherence to this dimension.

The median of the Performance dimension indicates that 50% of the dioceses reached 30 points or 75% of the requirements for the continuity of business and organizational operations and the long-term sustainability of human and environmental resources. This score also reveals a high adherence by dioceses to this dimension. The highlights are eight Brazilian dioceses (DBr1, DBr2, DBr3, DBr10, DBr13, DBr17, DBr22, and DBr32) that reached 40 points, or 100% adherence to that dimension. The Brazilian dioceses (DBr39, DBr52, DBr54, and DBr48) and an Italian one (DIT25) have the worst performance, with 19 points or 48% adherence to the requirements for organizational continuity.

In summary, from the perspective of the general median for each dimension of the Diocesan Corporate Governance Index (DCGI), the highest adherence (75%) is in the mechanisms of corporate governance (compliance with the goals and objectives planned through the provision of services to people and institutions) and performance (concern with the continuity of religious activities). Next are disclosure practices regarding accountability for the results of activities, with 54% adherence. Finally, the lowest adherence level is the corporate governance practice compliance with civil and canonical regulatory standards, with 44% adherence.

10.5 Discussion

10.5.1 The Diocese Corporate Governance Index (DCGI) Proposal

By discussing the level of adherence to corporate governance practices in the dioceses of Latin rite—Brazil, Spain, and Italy—through the proposition of the Diocesan Corporate Governance Index (DCGI), the results confirm findings from Inauen et al. (2010), López-Martínez and Fernández-Fernández (2015) and Pfang (2015), concerning the adoption of mechanism, compliance, and performance practices in religious and sports organizations. These findings were verified, even considering that dioceses have different adherence levels to these practices in the three countries studied.

The results also strengthen the arguments that the principles of well-consolidated corporate governance practices are relevant and applicable to dioceses' management, contradicting some mistaken and contrary views (Pfang, 2015). Such views concern that the Catholic Church is not a 'company' in the sense of a commercial organization; therefore, it should not be viewed administratively in the same way as for-profit companies (Simpson & Lane, 2012). Figure 10.2 shows the Church's conventional view of corporate governance.

In this perspective, what is accepted as good corporate governance practice in the dioceses, originating from Canon Law (João Paulo II, 1983), is subject to some criticisms, many of them justified by the context of forty years ago, when the Code of Canon Law was enacted. However, considering the results of this research, some considerations are necessary. One is that this governance does not guide curia processes. Another is that power is concentrated exclusively on the diocesan Bishop, and the last is the silence that hangs over the issues of responsibility and disclosure.

Given the study results and the identification of the conventional model, the study proposes a new ecclesiastical model of corporate governance, considering the practical implications, technological innovations, recommendations of acceptable practices, for example, 'do or explain', Moreover, that it is leveraged by the dimension of disclosure: transparency and accountability. However, it is emphasized that it is

Corporate governance practices	Description	Evaluation measures
Mechanism	Diocesan curia's offices, Council of Priests, College of Consultants.	Frequency of meetings, Compliance with the Code of Canon Law, and Form for the Five-Year Report.
Compliance	The Bishop's triple occupation of teaching, sanctifying and governing.	Nature of the discussions, decisions related to teaching, sanctification and governance
Performance	The Bishop's pastoral ministry in terms of sacraments, participation in events and the lay population.	Statistical data on the reception of the sacraments and the participation of the Catholic population

Fig. 10.2 Traditional Ecclesiastical Corporate Governance Model (*Source* Pfang [2015]; Adapted by the authors)

necessary to be careful not to use any model just because it is new. The Diocesan Corporate Governance Index (DCGI) seeks to minimize imperfections, focuses on the dioceses' corporate governance, not asking the bishops to leave his spiritual mission, but proposing them higher effectiveness in their government.

Among its four dimensions, the first is disclosure. Since disclosure is a constituent element of good corporate governance, the open publication of data will allow comparisons and promotion of accountability. In addition to producing efficient implications, such accessibility would allow comparisons between dioceses, favor the development of matrices, benchmarks, and acceptable practices. The research found, mainly in Italy and Spain's dioceses, probably due to specific legislation, a considerable adherence to this dimension.

In this scenario, it is necessary to expand the horizon of what accountability is. There is a mistake when this act is understood only in the scope of finance since accountability goes beyond financial tables. It is more in the field of sharing, expanding the church mural. It is to present, using the most current means and proper communication, the choices, the actions planned and executed, and sharing the dreams, making the interested parties feel that there are life and backbreaking work and competence in the organization.

Besides, the Catholic Church since the Counter-reform, passing through the Assembly of the Second Vatican Council, from 1962 to 1965, a priori, continues in the process of continuous innovation, when Pope St. John XXIII proposed to 'open the windows' of the Catholic Church to the contemporary world. He said at the time: 'I will open the window of the Church so that we can see what happens outside and so that the world can see what happens in our home'. There is the possibility of celebrating liturgies in each country's vernacular language and not only in Latin. Therefore, the disclosure dimension proposed by the Diocesan Corporate Governance Index (DCGI) is part of this process of 'opening the windows'.

The second dimension of the Diocesan Corporate Governance Index (DCGI) is that of conformity/compliance. The survey results also suggest the need to redefine the functions and relevance of some offices of the diocesan curia. The observation relates to the need for formal processes, such as regulatory inventory, code of conduct, audit committee or internal auditors, and arbitration status to resolve institutional conflicts. Thus, dioceses will be more qualified when it comes to preventing, detecting problems, and providing solutions.

The third is the mechanism. In this field, another practical implication is the need for specific training for the activities most exposed to risks (IBGC, 2017), to enable people to perform 'these activities more safely, including practical situations, case studies, and guidelines on how to resolve any dilemmas' (IBGC, 2017, p. 35). Without causing losses and damages to the institution. Finally, it is recommended that the diocese submit 'its compliance policies and procedures to an independent evaluation process, carried out by third parties, by the board of directors, internal audit or another supervisory and control body' (IBGC, 2017, p. 35). Such training will help bishops and curia members to operate effectively and consistently as a team.

Finally, the performance dimension. It is important to remember that the diocese and the bishop's *raison d'être* is spiritual, according to canons 771 and 761 of the

CIC, precisely, the salvation of souls. From it derives all its authority, the diocesan curia, the necessary administrative apparatus. All of them subordinate to this spiritual purpose, as can. 145 of the CIC. Ultimately, performance goes through a spiritual end. This goal justifies the implementation of new models of corporate governance that lead the Catholic Church to perform the realities that ensure its continuity—especially in the initiation into the Christian life and in the clergy's formation and life.

10.5.1.1 Accountability and Sanctions

For codes to be effective and long-lasting, there ought to be accountability and sanctions for non-compliance. Liability for acts of administration performed without the necessary requirements is regulated by can. 1281, §3 of the 1983 Code of Canon Law, and foresees two hypotheses:

a. When the act performed is invalid, it is the administrator himself and not the juridical person that must answer for the consequences and damages caused. This principle finds an exception when the invalid act derives an economic benefit and enrichment for the legal entity itself. On the other hand, the major problems in this area are related to the relationship with civil jurisdictions that can make a different legal evaluation of invalid acts and not recognize any value to elements considered in canon law, such as the lack of necessary licenses, that risk not being enforceable in court.
b. The second case refers to illegitimate but valid acts, in the presence of which the legal entity is patrimonially liable but retains the right to act or appeal against the administrators who by their acts have caused damage.

It should be noted that, apart from these norms which refer to the irregular acts committed, there is no systematic reconstruction of the responsibility of administrators. In this regard, *F.R. Aznar Gil in La administración de los bienes temporales de la Iglesia*, p. 374, insists on the fact that the legislation ought to better specify the cases of responsibility, underlining the need to establish exactly the cases in which the administrators will be responsible, both from the patrimonial point of view and in other acts of irresponsibility.

In the case of damage to the assets they administered, caused by the failure to comply with the duties inherent to the exercise of the administration, the liability of the administrators with respect to the legal entity is of a contractual nature. In this sense, the Rotal jurisprudence has ruled that the liability of the administrators in the development of their activity is not to be considered of Aquilian nature but is regulated by the norms relative to the position assumed (Cf. coram Many, decision of May 27, 1913, in RRDec., vol. V, p. 344. n. 14). An important consequence of this statement is that in the matter of guilt; the administrator is indeed liable for light fault, but not for very light fault, and in the rest the Code also provides that in administration one should proceed with the diligence of the good father of a family.

10.5.2 Theoretical and Practical Implications

The period of history that we live in has, among some peculiarities, rapid changes and, sometimes, discontinued. The new model of ecclesiastical governance—Diocesan Corporate Governance Index (DCGI) is significant and relevant, mainly because the social reality has changed, and the management models promulgated in the Code of Canon Law—mechanism, compliance, and performance—date back to 1983, as discussed by Pfang (2015). Furthermore, these practices are insufficient to cover the demands of the current context. Since the 1980s, there has been more volatility, uncertainty, complexity, and ambiguity in the world, requiring managers to make better decisions regarding the allocation of scarce resources to improve organizational performance (Bennett & Lemoine, 2014). As well as significant challenges to be faced in co-responsibility actions with other social actors (George et al., 2016).

Besides, in general, third-sector organizations have been charged by their stakeholders for better performance and greater accountability for the results (accountability) (Acar et al., 2012; Greiling & Stötzer, 2016; Grizzle & Sloan, 2016). Thus, in the connected and globalized world, with the technologies available, even at low cost, not adding disclosure (accountability for results and transparency) can be considered an omission and lack of connection with this reality. Adding disclosure dimension to the diocese's corporate governance model places decisions on how and when to put such a model into practice and exclude decisions on its feasibility and implementation. It is a proposal that complements the governance models in force by religious organizations.

The Diocesan Corporate Governance Index (DCGI) proposal shows that the practices already used successfully in the business world with a financial profit approach can be useful for religious and hierarchical organizations such as the Catholic Church. Therefore, it is necessary to overcome some barriers, such as the limited view that something created and developed in the market's logic cannot 'mix' with spiritual purposes and without financial profit. It is about promoting discussions focused on the search for excellence in its processes and results. In other words, to adopt practices of 'processes, customs, policies, laws, and institutions affecting the way an organization is directed and controlled' (Inauen et al., 2015, p. 764) more effectively.

The article develops a coherent, objective, and transparent statistical model, which can be applied on a worldwide scale, capable of seeking evidence of corporate governance in complex organizations, in terms of singularities—contingency (time and place) or paradox (conservative and progressive), as is the case with organizations with religious women. Although the study's object is the dioceses of the Roman Catholic Apostolic Church, this model can be applied without restrictions in other religious organizations, whatever the professed faith. Even in the face of idiosyncrasies in the various Christian (Protestant) communities outside the scope of the Catholic Church—Baptist, Methodist, Episcopal, Protestant, Presbyterian, Lutheran, and Pentecostal churches. There is evidence of the model's dimensions in its structures, such as fiscal supervision and internal controls (Elson et al., 2007).

In the Catholic Church, particularly, if the model is applied, crises such as that of the Sanctuary Basilica of the Divine Father Eternal, in Brazil, could be avoided with the governance practices of disclosure, mechanism, and compliance. In Italy, facing the Vatican Bank accusations, Pope Francis, in August 2019, amended the statutes to allow an external audit. That is, it applied the dimension of the mechanism and compliance. In Spain, a corruption scandal like the one in Borja parish would be history to illustrate what it was like and how it is today.

10.5.3 Limitations and Future Studies

The study is limited by the number of respondents, although the rate of return was 18.6%. Could this limitation be due to the culture of not practicing disclosure—transparency, accountability, accountability for results, disclosure of information to internal and external stakeholders—out of moral and not a legal obligation? The Code of Good Corporate Governance Practices is relatively recent. It is still unclear to what extent its principles are known and incorporated in non-profit organizations' realities, particularly in dioceses. Therefore, for the adequate development of a corporate governance structure aimed at religious organizations and based on the sector's etiology, further studies and debates are necessary.

Hence the suggestion of new studies, research in dioceses of other rites, in different religions and male and female religious orders/congregations, and the use of multivariate analysis as data treatment and examination. It may also be useful to compare the corporate governance practices proposed in this study with other non-governmental organizations in those countries. Also, the Catholic Church has significantly expanded in Africa and Asia. Therefore, future studies can incorporate data from these parts of the world. Finally, to mitigate the limitation related to accountabilities, future studies can test the index while experimenting on sanctions for violations.

References

Acar, M., Guo, C., & Yang, K. (2012). Accountability in voluntary partnerships: To whom and for what? *Public Organization Review, 12*(2), 157–174.

Almeida, M. A., & Zagaris, B. (2015). Political capture in the Petrobras corruption scandal: The sad tale of an oil giant. *The Fletcher Forum of World Affairs, 39*(2010), 87–99. https://static1.squarespace.com/static/579fc2ad725e253a86230610/t/57ec6f6fc534a5e78b120d0a/1475112816226/39–2_14AlmeidaZagaris.pdf

Ávila, L. A. C., & Bertero, C. O. (2016). Governança no terceiro setor: Um estudo de caso em uma fundação de apoio universitário. *Revista Brasileira De Gestao De Negocios, 18*(59), 125–144. https://doi.org/10.7819/rbgn.v18i59.2107

Barraket, J., & Yousefpour, N. (2013). Evaluation and social impact measurement amongst small to medium social enterprises: Process, purpose and value. *Australian Journal of Public Administration, 72*(4), 447–458.

Barth, T. (2010). Crisis management in the Catholic Church: Lessons for public administrators. *Public Administration Review, 70*(5), 780–791.

Battisti, D. A. (2010). A Visita "Ad Limina Apostolorum". *CNBB Igreja Católica Apostólica Roma* (pp. 1–3). https://www.cnbb.org.br/a-visita-ad-limina-apostolorum/

Bazanini, R., & Machado Junior, C. (2018). O Mercado como Religião: A Dinâmica da Rede de. *Brazilian Business Review, 15*(3), 262–283.

Beattie, V., McInnes, B., & Fearnley, S. (2004). A methodology for analysing and evaluating narratives in annual reports: A comprehensive descriptive profile and metrics for disclosure quality attributes. *Accounting Forum, 28*(3), 205–236. https://doi.org/10.1016/j.accfor.2004.07.001

Bennett, N., & Lemoine, G. J. (2014). What a difference a word makes: Understanding threats to performance in a VUCA world. *Business Horizons, 57*(3), 311–317.

Benjamin, L. M., & Campbell, D. C. (2015). Non-profit performance: Accounting for the agency of clients. *Non-Profit and Voluntary Sector Quarterly, 44*(5), 988–1006.

Borzaga, C., & Fazzi, L. (2014). Civil society, third sector, and healthcare: the case of social cooperatives in Italy. *Social Science and Medicine, 123*, 234–241.

Cavalcante, H. da S. (2018). Vademecum para os Bispos. In E. Juspodium (Org.), *Sociedade Brasileira de Canonistas* (4° ed). Salvador.

CEE. (2019). Conferencia Episcopal Espanhola - CEE. *conferenciaepiscopal.es.*

CEI. (2019). Conferencia Episcopal Italiana - CEI. *chiesacattolica.it.*

CFC. Conselho Federal de Contabilidade. (2009). *Resolução Conselho Federal de Contabilidade n. 1.185, de 15 de setembro de 2009.* Aprova a NBC TG 26 (NBC T 19.27) - Apresentação das Demonstrações Contábeis. https://www2.cfc.org.br/sisweb/sre/detalhes_sre.aspx?codigo=2009/001185

CFC. Conselho Federal de Contabilidade. (2012). *Interpretação Técnica Geral (ITG) 2002. Resolução nº 1.409, de 27 de setembro de 2012. – Entidade sem finalidade de lucros.* https://www2.cfc.org.br/sisweb/sre/detalhes_sre.aspx?codigo=2012/001409

Chiu, W. H. (2016). Political governance and economic growth: Evidence from emerging economies. *International Journal of Economic Perspectives, 10*(4), 376–394.

Cenfetelli, R. T., & Bassellier, G. (2009). Interpretation of formative measurement in information systems research. *MIS Quarterly, 33*(4), 689–707.

CNBB. (2019). Conferencia Nacional dos Bispos do BrasiL - CNBB. *cnbb.org.br.*

CNMV. (2016). Guía técnica de buenas prácticas para la aplicación del principio «cumplir o explicar». *Comisión Nacional del Mercado de Valores - CNMV* (pp. 1–14).

Coy, D., & Dixon, K. (2004). The public accountability index: Crafting a parametric disclosure index for annual reports. *British Accounting Review, 36*(1), 79–106. https://doi.org/10.1016/j.bar.2003.10.003

Downes, C. M. (2015). Appointing chapter 11 trustees in reorganizations of religious institutions. *Virginia Law Review, 101*, 2225–2262.

Elson, R. J., O'Callaghan, S., & Walker, J. P. (2007). Corporate governance in religious organizations: A study of current practices in the local Church. *Academy of Accounting and Financial Studies Journal, 11*(1), 121–131.

Europa Press. El párroco de Borja defiende su inocencia y agradece el apoyo de los vecinos del municipio. (2013). El Mundo, December, 20th. https://www.elmundo.es/espana/2013/12/02/529c707c0ab74081628b45a4.html

Faller, C. M., & Zu Knyphausen-Aufseß, D. (2016). Does equity ownership matter for corporate social responsibility? A literature review of theories and recent empirical findings. *Journal of Business Ethics, 1–26*. https://doi.org/10.1007/s10551-016-3122-x

George, G., Howard-Grenville, J., Joshi, A., & Tihanyi, L. (2016). Understanding and tackling societal grand challenges through management research. *Academy of Management Journal, 59*(6), 1880–1895. https://doi.org/10.5465/amj.2016.4007

Giumbelli, E. (2011). O Acordo Brasil-Santa Sé e as Relações entre Estado, Sociedade e Religião. *Ciencias Sociales y Religión, 13*(14), 119–143. https://doi.org/10.22456/1982-2650.19321

Gomes, M. S. (2009). A visita "Ad Limina Apostolorum". Recuperado de Sociedade Brasileira de Canonistas website: http://www.infosbc.org.br/site/artigos/616-a-visita-ad-limina-apostolorum

Greiling, D., Stötzer, S., & Stoetzer, S. (2016). Accountability reporting in Austrian nonprofit organizations-more than a compliance instrument?. *Public Administration Quarterly*, 256–287.

Grizzle, C., & Sloan, M. F. (2016). Assessing changing accountability structures created by emerging equity markets in the non-profit sector. *Public Administration Quarterly*, 387–408.

Hair, J. F., Jr., Babin, B., Money, A. H., & Samuel, P. (2003). *Essentials of business research methods*. John Wiley.

Heinrich, F., & Brown, A. J. (2017). Measuring accountability performance and its relevance for anti-corruption: Introducing a new integrity system-based measure. *Crime, Law and Social Change, 68*(3), 359–381.

IBCG. Instituto Brasileiro de Governança Corporativa. (2017). O Papel do Conselho de Administração na Estratégia das Organizações (pp. 1–56).

IBGC. Instituto Brasileiro de Governança Corporativa. (2015). Códi*go das melhores práticas de governança corporativa*. IBGC. 108p. https://conhecimento.ibgc.org.br/Paginas/Publicacao.aspx?PubId=21138

IBGC, I. B. de G. C. (2014). *Guia das Melhores Práticas de Governança para Institutos e Fundações Empresariais*. www.gife.org.br

Inauen, E., Osterloh, M., Frey, B. S., & Homberg, F. (2015). How a multiple orientation of control reduces governance failures: A focus on monastic auditing. *Journal of Management and Governance, 19*, 763–796. https://doi.org/10.1007/s10997-014-9292-y

Inauen, E., Rost, K., Osterloh, M., & Frey, B. S. (2010). Back to the future–A monastic perspective on corporate governance. *Management Revue, 20*(2), 158–175.

Irene, B., Marika, A., Giovanni, A., & Mario, C. (2016). Indicators and metrics for social business: A review of current approaches. *Journal of Social Entrepreneurship, 7*(1), 1–24.

João Paulo II, P. (1983). *Constituição Apostólica "Sacrae Disciplinae Leges" de Promulgação do Código de Direito Canônico* (Loyola, Org.).

Kissler, L., & Heidemann, F. G. (2006). Governança pública: Novo modelo regulatório para as relações entre Estado, mercado e sociedade? *Revista De Administração Pública, 40*(3), 479–499.

Kroeger, A., & Weber, C. (2014). Developing a conceptual framework for comparing social value creation. *Academy of Management Review, 39*(4), 513–540.

Ladeur, K. H. (2017). Globalization and public governance—A contradiction? *Public governance in the age of globalization* (pp. 1–22). Routledge.

Lopes, A. B., & Walker, M. (2008). Firm-level incentives and the informativeness of accounting reports: An experiment in Brazil. In: *AAA Annual Meeting 2008—An International Meeting of the American Accounting Association*, Anaheim, California. https://ssrn.com/abstract=1095781 or https://doi.org/10.2139/ssrn.1095781

López-Martínez, R., & Fernández-Fernández, J. L. (2015). Responsabilidad social corporativa y buen gobierno en los clubes de fútbol españoles. *Universia Business Review, 46*, 38–53.

Lugoboni, L. F., Alencar, E. A. de B., Zittei, M. V. M., & Chirotto, A. R. (2018). Alinhamento entre Planejamento Estratégico e Sistema de Informação Gerencial: Estudo em Empresas do Terceiro Setor com Atividade Hoteleira. *PODIUM Sport, Leisure and Tourism Review, 7*(1), 79–105.https://doi.org/10.5585/podium.v7i1.244

Mariani, G., Bernini, F., & Panaro, D. (2013). The corporate governance drivers: What relations with performance and risk? Empirical evidence from Italian context. *Corporate Ownership and Control, 11*(1 K), 766–780. https://doi.org/10.22495/cocv10i4c5art7

Melão, N. F., Maria Guia, S., & Amorim, M. (2017). Quality Management and Excellence in the third sector: Examining European Quality in Social Services (EQUASS) in non-profit social services. *Total Quality Management and Business Excellence, 28*(7–8), 840–857.

Messi, J. M. (2010). VISITA AD LMINA APOSTOLORUM por Dom João Maria Messi, OSM. *INOSBC* (p. 3). http://www.infosbc.org.br/site/75-o-povo-de-deus-i-da-educacao-fieis-leigos/ii-ii-seccao-igrejas-particulares-e-sua-autorida/1245-visita-ad-lmina-apostolorum-por-dom-joao-maria-messi-osm

Millar, C. C., Eldomiaty, T. I., Choi, C. J., & Hilton, B. (2005). Corporate governance and institutional transparency in emerging markets. *Journal of Business Ethics, 59*(1–2), 163–174.

Murtinho, M. N., de Castilho, F. M. B. M., & Urdan, A. T. (2018). Declínio Organizacional e Contexto em Organização Religiosa: A Igreja Universal do Reino de Deus. *Revista Ibero-Americana De Estratégia, 17*(02), 93–107. https://doi.org/10.5585/riae.v17i2.2595

Oliveira, R. (2020). Conheça a trajetória do padre Robson e entenda a investigação do MP sobre a compra de imóveis de luxo com dinheiro de fiéis. *G1 GO*, August 27. https://g1.globo.com/go/goias/noticia/2020/08/27/entenda-a-ascensao-de-padre-robson-na-igreja-as-movimentacoes-milionarias-de-sua-entidade-e-o-escandalo-de-desvio-de-dinheiro.ghtml

Oliveira, M. C., Ponte, V. M. R., De Luca, M. M. M., de Oliveira, O. V., Aragão, L. A., & Geleilate, J. M. G. (2013). Práticas de governança corporativa adotadas por companhias fechadas brasileiras e alinhamento às demandas do mercado de capitais. *BASE - Revista De Administração e Contabilidade Da Unisinos, 10*(3), 196–209. https://doi.org/10.4013/base.2013.103.01

Payer-Langthaler, S., & Hiebl, R. W. M. (2013). Towards a definition of performance for religious organizations and beyond a case of Benedictine abbeys. *Qualitative Research in Accounting and Management, 10*(3/4), 213–233. https://doi.org/10.1108/QRAM-08-2012-0033

Pfang, R. (2015). Management in the catholic church: Corporate governance. *Journal of Management, Spirituality and Religion, 12*(1), 38–58. https://doi.org/10.1080/14766086.2014.933708

Pless, N. M., Maak, T., & Stahl, G. K. (2017). Promoting corporate social responsibility and sustainable development through management development: What can be learned from international service learning programs? *Human Resource Management, 51*(6), 873–903. https://doi.org/10.1002/hrm

Rezende, A. J., & Dalmácio, F. Z. (2015). Práticas de Governança Corporativa e Indicadores de Performance dos Clubes de Futebol: Uma Análise das Relações Estruturais. *Contabilidade, Gestão e Governança, 18*(3), 105–125.

Rezende, A. J., Dalmacio, F. Z., & Facure, C. E. F. (2010). Practice of corporate governance in football clubs. *International Journal of Economics and Accounting, 1*(4), 410. https://doi.org/10.1504/ijea.2010.037579

Rost, K., Inauen, E., Osterloh, M., & Frey, B. S. (2010). The corporate governance of Benedictine abbeys: What can stock corporations learn from monasteries? *Journal of Management History, 16*(1), 90–115. https://doi.org/10.1108/17511341011008331

Schmidt, P. (2014). NGOs as a framework for an education in and through music: Is the third sector viable? *International Journal of Music Education, 32*(1), 31–52.

Silva, A. L. C., & Leal, R. P. C. (2005). Corporate governance index, firm valuation and performance in Brazil. *Brazilian Review of Finance, 3*(1), 1–18.

Silveira, A. D. M. (2005). Governança corporativa e estrutura de propriedade: determinantes e relação com o desempenho das empresas no Brasil. Saint Paul Institute of Finance.

Silveira, A. D. M., Perobelli, F. F. C., & Barros, L. A. B. D. C. (2008). Governança corporativa e os determinantes da estrutura de capital: Evidências empíricas no Brasil. *Revista De Administração Contemporânea, 12*(3), 763–788.

Simpson, P., & Lane, C. (2012). Complexity and change management: Analyzing church leaders' narratives. *Journal of Organizational Change Management, 25*(2), 283–296.

Tihanyi, L., Graffin, S., & George, G. (2015). Rethinking governance in management research. *Academy of Management Journal, 57*(6), 1535–1543.

Vatican News. (2019a). A contribuição das religiões para os objetivos do desenvolvimento sustentável. *Vatican News*, 3–7.

Vatican News. (2019b). O Papa Francisco renova os Estatutos do IOR: auditor externo verifica os balanços. Vatican News, online. https://www.vaticannews.va/pt/papa/news/2019b-08/papa-francisco-renova-os-estatutos-do-ior-auditor-externo.htmlVatican

Vatican, S. di S. (2018). Anuario Pontificio 2018 y Annuarium Statisticum Ecclesiae 2016. *L'Osservatore Romano.*

Vaticano, C. E. (1962). *Constituição Dogmática Lumen Gentium: sobre a Igreja.* (Vozes, Org.). Petrópolis.

Verdú, D. (2019). O buraco negro das finanças do Vaticano. *El País*, November 4. https://brasil.elpais.com/brasil/2019/11/02/internacional/1572705299_833356.html

Windrum, P. (2014). Third sector organizations and the co-production of health innovations. *Management Decision, 52*(6), 1046–1056.

Wolff, J. (2018). Political incorporation in measures of democracy: A missing dimension (and the case of Bolivia). *Democratization, 25*(4), 692–708.

Yong, G., & Wenhao, C. (2012). Developing a city governance index: Based on surveys in five major Chinese cities. *Social Indicators Research, 109*, 305–316.

Chapter 11
Relationships Among the New Facet Leadership Style, Quality of Work Life, Satisfaction and Performance in the Secondary Schools of Mauritius

Subrun Veerunjaysingh

Abstract Education can be well-thought-out to be the key to a successful future of any nation (Subrun and Subrun, International Journal of Learning, Teaching and Educational Research, Special Issue 13:14–19, 2015). Within the pandemic situations, where the whole world is suffering, various governments have kneeled down due to economic and health deteriorating features, the education system has still remained a priority for many of them. The Mauritian Government has been providing a caring environment and has made the provision of facilities to enhance the teaching and learning process in the school (Subrun and Subrun, International Journal of Learning, Teaching and Educational Research, Special Issue 13:41–48, 201), even in this harsh situation of the pandemic. The leadership style of the school leaders, quality of work life and satisfaction have been the key factors in determining the performance. The aim of the study is to understand the relationship among the New Facet Leadership Style (Subrun and Subrun, Journal of Business and Social Sciences 4:1–9, 2022; Subrun and Subrun, International Journal of Engineering Applied Sciences and Technology 7:67–72, 2022; Subrun and Subrun, Impact of the New Facet Leadership Style on job satisfaction in the Secondary Schools of Mauritius, Journal of Business and Social Sciences, Vol. 7, 2022), quality of work life, satisfaction and performance in the secondary schools of Mauritius. 500 educationists from the private and state secondary schools of Mauritius took part in the questionnaire filing process. The study investigated the impact of the New Facet Leadership Style, quality of work life in relation to job satisfaction and performance in the secondary schools in Mauritius through the correlation analysis and multiple regression analysis using SPSS. Data has been gathered from rectors and educators of the secondary schools. The study provides new heights for other researchers to further explore the notion of quality of work life, job satisfaction and New Facet Leadership Style at the National and International levels.

Keywords New Facet Leadership Style · Education · Job satisfaction · Quality of work life · Performance · Secondary schools · Leadership

---

S. Veerunjaysingh (✉)
Université Des Mascareignes, Rose-Hill, Mauritius

D. Crowther and S. Seifi (eds.), *Preparing for a Sustainable Future*,
Approaches to Global Sustainability, Markets, and Governance,
https://doi.org/10.1007/978-981-99-2456-1_11

11.1 Introduction

Employees' satisfaction is appreciably correlated and is influenced by leadership style (Barnett, 2018; Cansoy, 2019; Mahzan & Nordin, 2021) implemented by leaders in an employee-oriented dimension. Subsequently, it may be claimed that the organizational success in achieving the goals and objectives of the institution depends on the leadership style adopted by the leader. Job satisfaction is the extent to which a person has positive and negative emotional state about a profession, other employees and work milieu (Schermerhorn et al., 2011). Moreover, job satisfaction can be described as an emotion, a feeling, an attitude and a matter of perception. It begins from the employee's appraisal of experience at the job. It involves likes and dislikes as well as needs and wants both internal and external. Thus, the job satisfaction may be either positively or negatively related to the quality of work life, attitude, self-concept, self-efficacy and leadership style of the leaders. Many researchers have demonstrated that there is a positive relationship connecting leadership and job satisfaction (Berson & Linton, 2005; Mosadeghrad & Larson, 2002; Seo et al., 2004). With the same notion, Zeithaml and Bitner (2000) pointed out that employees' job satisfaction determines their performance. Educational leaders' managerial actions that encourage involvement and are flexible, sharing leadership at school, displaying individual-oriented and supportive leadership actions were discovered to augment educators' job satisfaction (Cansoy, 2019). According to Mancha and Yoder (2015), there is a positive influence of leadership style on job satisfaction. With the same notion, Robbins and Coulter (2016) claimed that job satisfaction remarkably impacted educators' intention to stay. Based on factors like job satisfaction and the quality of work life, the staff members of the institution have an impact on the decision of the intent to stay in the institution. Moreover, the above notion can be further supported by the fact that the study of Robbins and Coulter (2016) with regards to the influence of workers' job satisfaction on the turnover objective, revealed that staff would try to find other job prospects and evaluate the advantages and disadvantages of shifting to other jobs to make an intention, and then to stay in the present position or leave the job.

11.2 Aims and Objectives of the Study

The study's aim is to explore the impact of New Facet Leadership Style on the quality of work life, satisfaction and performance in secondary schools in Mauritius. It intends to determine whether the New Facet Leadership Style affects the quality of work life, satisfaction and the secondary school performance.

11.3 The Need for the New Facet Leadership Style

The New Facet Leadership Style has been adapted from Subrun and Subrun (2022a, 2022b, 2022c) study. The New Facet Leadership Style is a completely new bold leadership style that comprises of various leadership style. Performance of the educational institutions rests on good administration and improved leadership styles of the educational leaders, set as performance features that every educational leader must own (Gultom & Fibriasari, 2021) and leadership is vital. It can be further claimed that integrating ideals in handling an educational institution is extremely important to have a sense of balance to play the role of educational leaders as leaders and managers below one roof (Vikaraman et al., 2021) as leadership is the key of the success of the educational institutions. Today with the advent of urbanization and a change in mentality and due to the awareness through media and internet, dealing with all types of unpredictable circumstances has become challenging. The application of a unique style of leadership do not cater for the need of the subordinates and handling unforeseen circumstances is no more functional in the institutions. Thus, the performance of educational institutions can be enhanced by applying the appropriate leadership styles (Kirby et al., 2021) at the right instance. School leadership is very important and it is required to be developed in educational management (Gultom & Fibriasari, 2021) as the education system is always dynamic and it needs to adapt to survive in this everchanging world. Leadership is an ancient occurrence that shapes quite a lot of subordinates' results (Mishra & Pandey, 2018) and it can be used to uplift the institutions. The educational managers are inspired to organize more leadership training programs for school leaders, and make sure that educators are sufficiently inspired to boost the academic performance of pupils (Adarkwah & Zeyuan, 2020) so that they can excel in this competitive world. Leadership approaches of the leaders cover various extents of leadership style (Haile, 2017) as a unique style cannot cater for all types of situations. Since, leadership has no customary form or manner, and diverse leaders utilize numerous methods and skills (Ho & Lin, 2015), the need for a better leadership style in almost all types of institutions have been felt. It can be noted that Gandolfi and Stone (2018) highlighted that the manner a leader selects to act, or in more academic jargon the way a leader accesses a collection of styles, influences a number of stakeholders profoundly. Subsequently, Fiaz et al. (2017) claimed that leadership supports institutions to be more fruitful and profitable, but the degree of achievement is ruled by the style of the subsequent milieu generated for staff to work well. Thus, leaders do not strictly use one or another approach of leadership, but are somewhere on a continuum ranging from extremely positive to extremely negative (Clark & Clark, 2000) to counteract the unforeseen situations. Ultimately, the unbiased link and the data gathered confirmed the blend of leadership styles do have an impact on job satisfaction as workforces reach complete willingness, which affirmed its unidirectional capability and presence (Matthews, Daigle & Houston, 2018). Thus, to heighten leadership efficacy, a leader should employ more than one leadership style, change appropriate leadership styles according to the development stages and goals of the institution to encourage followers' potential and to attain the institutional objectives (Li & Wu, 2018). In the same vein, Yao and Huang (2018) regarded leadership as guiding and instructing people or teams, in a particular situation, to smoothly realize a goal. Consequently, it may be noted that all kinds of leadership were efficient, depending upon circumstances (Fieldler, 1967). Accordingly, Lewin (1945) a psychologist led a group of researchers to identify different styles of leadership. The research revealed that there are more specific types of leadership and this study was very influential as it established three major leadership styles. Subsequently, Lewin came forward with a new leadership style termed as the Kurt Lewin Leadership Style. The Kurt Lewin Leadership Style comprised of three different leadership styles. The Authoritarian, Participative and Delegative leadership style were the three major leadership styles which were used. Thus, for the study, the New Facet Leadership Style has been developed following the same pathway of Kurt Lewin (1945) model of leadership. Kurt Lewin had made use of the directive leadership or participative leadership along with democratic leadership or autocratic leadership style to establish the Kurt Lewin Leadership Style. In the same way the New Facet Leadership Style encompasses the contingency leadership, instructional leadership, managerial leadership, moral leadership and participative leadership along with the transformational leadership styles to handle all types of situations and at the same time to generate a positive influence of the subordinates in terms of attitude, self-efficacy, quality of work life, satisfaction, self-concepts, work culture and performance. The New Facet Leadership Style is based on the Contingency Leadership Theory and the Transformational Leadership theory. A number of scholars have claimed through a number of studies that different types of leadership styles have worked best. For example: Cansoy (2019) concluded that educational leaders' managerial actions that encourage involvement and are flexible, sharing leadership at school and displaying individual-oriented and supportive leadership actions were discovered to augment educators' job satisfaction; Duyan and Yildiz (2020) concluded that transformational leadership and job satisfaction will not only provide individual growth of academic personnel, but also will give an inclusive performance of the academy such as academic publications, pupils satisfaction and teamwork among staffs; Donkor and Asante (2016) noted that if there is an insufficient time disbursed on instructional leadership by school leaders, this possibly will account for a meagre performance of both teachers and pupils or students in the basic school of study; Bronskhorst, Steijin & Vermeeren (2013) noted that transformational leadership characteristics are capable of influencing positively employee motivation, by dedicating their endeavours and determination to specific objectives and tasks; Bambale, Girei & Barwa (2017), noted that servant leadership is defined as serving others by working in the direction of their progress and happiness; Wu (2012) noted that moral leadership enhanced job performance since such actions improved the reliability of managers and sequentially boosts juniors' intrinsic enthusiasm as well as mental enablement, which ultimately geared towards the improvement of job performance; Chan (2019) noted that the participative leadership was constructively connected to workers' job commitment and work contentment; Nyberg, Holmberg, Bernin, Alderling, Akerblom, Widerzal-Bazyl, Magrin, Hasselhorn, Milcrarek, Dangelo, Denk, Westerlund & Theorell (2011) postulated that managerial leadership emphasized on positive leader features and deeds, the leaders addressed emotional state, attitudes, and gratification of the followers, enhancing self-esteem, enhancing interconnection, diminishing skirmish and so forth; Bellibas and Liu (2017) concluded that there was a noteworthy and constructive association among school leaders professed instructional leadership exercise and educators' self-efficiency in schoolroom administration, instruction and pupil engagement; Dipboye (2018) noted that the contingency theories of leadership arose from investigation representing the circumstances under which the different types of qualities and deeds of headship are related with gratification and effective presentation. Therefore, it may be inferred that a unique leadership style cannot be applied all the time. The transformational leadership style has been adopted because scholars like Zhang and Tan (2021) noted that leaders who employ the transformational leadership instigate all of their members to work in the direction of a communitarian vision by using situational decisions that are steered by contemplation, and by working namelessness. Transformational leadership can be related to various constructive individual, group and organizational results related to management, mindset and achievement (Arnold & Loughlin, 2013). Consequently, this particular leadership style could be exercised in the forms of directive leadership or participative leadership along with democratic leadership or autocratic leadership style (Avolio, 2011). Thus, transformational leadership theory can be integrated with all types of effective leadership styles. It was directly and indirectly constructively connected for group acquirements (Bouwmans et al., 2017) and the usage of transformational behaviours on a daily basis relation could support workers' insights about their work features and could grow sentimental attachment (Islam et al., 2018). In the same vein, Mahzan and Nordin (2021) concluded that the transformational leadership features demonstrate an imperative part in fostering and developing positive attitudes among educators to realize the marked aims and objectives. Transformational leadership must be embedded in the educational institutions, as it provides individual growth of academic personnel, but also gives an inclusive performance of the academy such as academic publications, pupils satisfaction and teamwork among staff (Duyan & Yildiz, 2020).

The success of educational leadership is mainly governed by the various leadership styles depending upon the situation, thus, for the advent of the New Facet Leadership Style, the following six main components of leadership: Instructional leadership, Moral leadership, Participative leadership, Managerial leadership and Contingency leadership along with the Transformational leadership were unified and integrated to cater for various specific circumstances. The amalgamation of the various leadership styles to cater for various situations may be further justified by the fact that Yun et al. (2006) argued that one type of headship is possible to succeed in specific circumstances, and in a diverse circumstance, another type of guidance would be more suitable and these concepts is from time to time termed as "situational" view, or as they termed it in their research as "contingency" view. Thus, the educational leaders must employ the appropriate leadership styles based upon the situation and the demand. At the same time, the leaders must employ a myriad of leadership styles (along with transformational leadership style in terms of the New Facet Leadership Style) to enhance workers' performance in an institution (Bambale et al., 2017), while ensuring that the leadership style positively affects the attitude of the staffs, their self-efficacy, quality of work life, satisfaction, participation in nonworking related activities, perception, self-concepts, work culture, learners' achievement and school performance. Thus, the present study intends to scrutinize the leadership style in terms of the New Facet Leadership Style based on the contingency leadership, instructional leadership, managerial leadership, moral leadership and participative leadership along with the transformational leadership in relation to the attitude of the staff, their self-efficacy, quality of work life, satisfaction, participation in nonworking related activities, perception, self-concepts, work culture, learners' achievement and school performance.

Moreover, the leadership style sets the tone of the organizational environment and natures the attitude and performance of the followers (Boyer-Davis, 2018). Therefore, a myriad of leadership styles in terms of contingency leadership, instructional leadership, managerial leadership, moral leadership, participative leadership along with the transformational leadership style may turned out of immense importance. Subsequently, the schools are not statics and a standard type of leadership style cannot be applied to handle the various types of situations. The New Facet Leadership Style comprises contingency leadership, instructional leadership, managerial leadership, moral leadership and participative leadership along with the transformational leadership styles to handle all types of situations and at the same time to generate a positive influence of the subordinates in terms of attitude, self-efficacy, quality of work life, satisfaction, perception, participation in nonworking related activities, self-concepts, work culture and performance. The New Facet Leadership Style is based on the Contingency Leadership Theory and the Transformational Leadership theory. Subsequently, the leaders usually work in line with the contingency theory (Fiedler, 1951; Northouse, 2004), as the leader's effectiveness depends on how well the leader's style matches the specific setting or situation and the transformational leadership theory (Bass, 1998, 2008) to consistently motivate, inspire and optimized supporters' effort to accomplish what could be beyond their reach but at the same time it also enhances the inspirational appeal of the leaders (Ghaus et al., 2017). Thus, the leaders must master the various leadership styles and they will be able to handle the various situation using these leadership styles: contingency leadership, instructional leadership, managerial leadership, moral leadership and participative leadership while incorporating the transformational leadership style termed as the New Facet Leadership Style, to cater for the needs of the subordinates and to handle all type of situations. With the same perception, Baptiste (2019) claimed that the actions of educational leaders greatly influence the involvement of the educators as well as the general performance of the educational institutions. Accordingly, the application of the right leadership style at the right time is of great importance. In the same context, it can be noted that the leaders had to adapt to the reality of the demands made by the increasingly complex and continually evolving role call for a different leadership style (Webb, 2005). Consequently, Yu (2016), further claimed that if managers offer further influential backing, attention and faith to the staff, this would generate a constructive cooperative connection among managers and workforces, and the inventiveness of staff would be more reinforced. Thus, leadership style must engross the good qualities of the leader that can be utilized to handle appropriately the day-to-day situation, which in turn will encourage and guide the subordinates towards an enhanced performance. Subsequently, the present study aims to scrutinize whether the leadership style in terms of the New Facet Leadership Style affects the quality of work life, work culture, attitude, self-efficacy, self-concept, satisfaction and the learners' achievement in terms of school performance in the Mauritians' secondary schools.

11.4 Satisfaction and Leadership

It is a mere fact that many subordinates opt to join other institutions even with inferior pay in the quest for job satisfaction. As a leader or a manager, if one is unable to meet their expectation, there is a high chance one will also not achieve high performance. Creating job satisfaction stays a challenge for many leaders. Subsequently, DeCarlo and Agarwal's (1999) study sustained that a positive correlation exists between job satisfaction and leadership. In the same line, both Tsai and Chang (2016) and Lee and Lam (2016) claimed that a higher value of leadership produces a higher level of employees' job satisfaction. Transformational leadership characteristics such as empowerment and clear vision, are habitually seen as essential elements for employee job satisfaction and commitment (Iverson & Roy, 1994) and this has been sustained by other authors like Sergiovanni and Corbally (1984) and Smith and Peterson (1988). Moreover, Petyer's (2004) study confirmed that organizational culture and leadership style are the essential organizational backgrounds of job satisfaction and commitment. With the same perception, Matthews, Daigle and Houston (2018) claimed that the unbiased link and the data gathered confirmed the blend of leadership styles do have an impact on job satisfaction as workforces reach complete willingness, which affirmed its unidirectional capability and presence. Job satisfaction is linked with the leadership style of the leaders, the appropriate leadership style empowers the subordinates to own the process and thus they perform extremely well since they are satisfied with the good ambience of the institution. However, if there is inappropriate usage of leadership style, a tense atmosphere prevails in the institution. The subordinates are in search of the minimal opportunity to shift to another job as they do not have an intent to stay since they are not satisfied. Consequently, the performance of the subordinates is not up to the desired level, since they are neither intrinsically nor extrinsically motivated. Thus, the aspect of job satisfaction is a crucial characteristic that has a great impact on the learners' achievement and school performance. Moreover, job satisfaction is to a great extent dependent on the type of leadership style (Lee & Lam, 2016; Tsai & Chang, 2016) the school leaders are exercising in the educational institutions. Subsequently, Baptiste (2019) claimed that educational leadership plays a noteworthy character in shaping the involvement of educators, the involvement of pupils and the global school climate. On the contrary, Cansoy (2019) noted that there was an adverse association between job satisfaction and laissez-faire leadership. However, Kouni et al. (2018) claimed that the educators have an impression of considerable satisfaction when the school leaders perform transformational leadership style. Subsequently, the latter further concluded that the demographic variables, the nature of the educational institution and the work experience, do not have an impact on the opinions of educators. Transformational leadership style was found to be closely related to a healthy school environment (Toprak et al., 2015). School culture might be effective in dropping undesirable behaviours of the educators concerning the institutional transformation (Atasoy, 2020). The utmost esteemed type of educational leadership is that one that is effectual and have an impact on a good performance at work (Gultom & Fibriasari, 2021). In view to enhance the quality of work life and job satisfaction, Adarkwah and Zeyuan (2020) noted that the educational managers are inspired to organize more leadership training programs for school leaders and make sure that educators are sufficiently inspired to boost the academic performance of pupils. The leadership styles affect the followers both attitudinally and behaviorally through job satisfaction, work performance, innovation, organizational commitment and so on (Alkahtani, 2016). The virtuous performance of the educational institutions rests on good administration and improved leadership styles of the educational leaders, set as performance features that every educational leader must own (Gultom & Fibriasari, 2021). Transformational leaders are capable of creating a positive organizational ambience, accomplishing goals more easily and boosting up the levels of job satisfaction and organizational commitment among the stakeholders as a result of motivating subordinates and paying close attention to them (Deluga & Souza, 1991). With the same perception, Demir (2020) claimed that the more educators' self-efficacy certainty is amplified, the more their job satisfaction, institutional obligation, enthusiasm and job participation augmented. The latter further claimed that self-efficacy certainty positively impacted educators' job participation through full mediation consequence of job satisfaction and motivation. Moreover, it can be noted that transformational leadership comportment among school managers had a noteworthy, direct and constructive influence on the quality work life perceptions of the educators and also had a noteworthy, indirect and positive effect on teachers' perceptions of quality of work life through their perceptions of organizational sustenance and organizational impartiality (Akar & Ustuner, 2019). The quality of work method is the name of the conversion from an establishment that considers the workforces imperative, positions the workforces into a central situation and esteems the workforces as the most significant part of the institution (Demir, 2011). Subsequently, the quality of work life leads the workforces to attain and institutional uniqueness and upsurges job performance, job satisfaction and professional initiatives of the workforces, declines the rate of absenteeism, intent to leave job and burnt-out stages (Kheirandish, 2009). Thus, job satisfaction can be described as the extent to which a person has a positive or negative emotional state about a profession, other employees and work milieu. Improved quality of work life has the capability to confirm a balance between professional, individual and societal life and this further augment job satisfaction and work obligation (Susillaningsih et al., 2021). Subsequently, it can be noted that some academics view job satisfaction as being constructive emotional reactions and attitudes a staff has towards their profession. In the same line, Rifadha and Sangarandeniya (2015) defines the term job satisfaction as a worker is said to be contented from his/her work when the latter senses well-being while working in the institution. Quality of work life is a sense of satisfaction that personnel have for their profession, colleagues and establishments which leads to the evolution and effectiveness of the institution (Heskett et al., 1997). With the same perception, Helmiatin (2014) claimed that in order to realize the institutional aims and objectives, a leader must be able to produce an enjoyable working atmosphere for the workers. The latter further noted that the crucial indicator of the performance enhancement is marked by the application of quality of work life. Accordingly, the present study aims to study the relationship between the leadership styles in terms of the New Facet Leadership Style of the school leaders in relation to quality of work life, work culture, self-efficacy, self-concept, attitude, satisfaction and school performance. Since leadership and motivation cannot be separated (Adarkwah & Zeyuan, 2020), leaders have a great influence on their followers. Job satisfaction affects the way people think about what is desirable, possible and necessary. Subsequently, it may be claimed that the satisfaction of the educators can be either positively or negatively affected by the leadership styles of the school leaders, which may be affected by quality of work life and eventually affect the overall performance of the educational institutions.

11.5 Quality of Work Life

According to Huzzard (2003) quality of work life is the humanization of work, enhancement of working circumstances, protection of personnel and democratization of working atmosphere. Even in this pandemic situation, the quality of work life in the educational institutions remains an important factor that affects the success and failure. The educators are important actors of the educational institutions and the performance of the school relies heavily on them. The educator's performance is the course of dealings between educators and pupils in the learning process, the interaction between the educational leader and educator and pupils in the learning process as well as between citizens/communities around neighbourhood educational institutions (Yuvitasari, 2018) and the leadership style of the leaders may affect the quality of life in the educational institutions. Management plays a larger part in supporting the achievement of the quality work life program, which is essential to endorse prosperity, both for the workers and the management (Secapramana & Nugroho, 2017). With the same perception, it can be noted that leadership styles have a direct impact on job satisfaction, institutional performance and workers quality of work life. The leadership styles are connected with enhanced quality of work life, as leadership lies on shared actions, where the team followers work together in whatever characters of authority and supremacy they may have, not based on position supremacy. Accordingly, to realize the institutional aims and objectives, the leader must be able to produce an enjoyable working atmosphere for the workers (Helmiatin, 2014) which may lead to satisfaction and enhanced performances.

11.6 Uniqueness of the New Facet Leadership Style

The New Facet Leadership Style is a new leadership style that demarcates itself from the other leadership style which have been criticized over time that there is no unique leadership style that work best in all situations. The New Facet Leadership Style is a new leadership style that tries to respond to the need of the leaders to govern the organizations effectively. It comprises of a selection of leadership style that remediates the gap in the organizations so that they can achieve new heights. Thus, the leaders must apply the New Facet Leadership Style judiciously to remediate the gap in the organizations. This can be supported by the fact that Gandolfi and Stone (2018) highlighted that the manner a leader selects to act, or in more academic jargon, the way a leader accesses a collection of styles, influences a number of stakeholders profoundly. Thus, the present New Facet Leadership Style, if applied properly by the leaders, it will be able to handle all types of unpredictable situations. The New Facet Leadership Style is a bold and complete leadership style that can create a substantial positive on the subordinates and the institutions. This leadership style has a substantial positive effect like transformational leadership, since it encompasses a number of leadership styles along with the transformational leadership style which enables the leaders to deal with various types of situations. In the same vein, the New Facet Leadership Style encourages the subordinates to perform better and thus creating an atmosphere where the subordinates work with professionalism. The New Facet Leadership Style can produce a positive quality of work life of the subordinates within the organization which in turn give rise to a conducive working environment. The conducive working environment creates satisfaction among the subordinates and thus the willingness among the subordinates to strive for a better performance occur. Moreover, the New Facet Leadership Style can be the future torchbearer of the institutions if the leaders in all types of institutions adopt the New Facet Leadership Style and applied it properly. Furthermore, the New Facet leadership Style comprises a number of leadership styles, thus the leaders must employ it judiciously to counteract the unforeseen situations so as to generate a better quality of work life which will lead towards job satisfaction and ultimately enhance the overall performance of the institutions to new heights. Thus, the New Facet Leadership Style may be applied based on the likelihood viewpoint. The New Facet Leadership Style comprises of flexible alterations to the prospects and encounters of an institutional institute's core and peripheral background. The New Facet Leadership Style acts as a guide for the leaders so that they can take effective decisions. The leaders of the institutions may handle the followers with the appropriate leadership skills through the New Facet Leadership Style by creating a positive impact on the quality of work life in terms of leadership skills, work culture, subordinates' attitude towards their job, occupational self-efficacy, subordinates' participation in non-work-related activities, self-concept and subordinates' perception about quality performance in the organization's. The New Facet Leadership Style establishes an efficient plan that comprises of overseeing and assessing instruction, motivating, inspiring, regulating aims and objectives and controlling advancement. Ultimately, to excel in this competitive world, the leaders must employ the New Facet Leadership Style which comprises of the contingency leadership style, instructional leadership style, managerial leadership style, moral leadership style and participative leadership style along with the transformational leadership style. Through the New Facet leadership Style, the leaders can accentuate positive features and deeds in the institutions. They can achieve the goal of the institution through the New Facet Leadership Style as it enhances the emotional state, attitudes, and gratification of the followers, work culture, enhances self-esteem, enhances interconnection, enhances the quality of work life, enhances self-efficacy, enhances self-concept, enhances satisfaction, enhances motivation, inspires the subordinates, empowers the subordinates, values the subordinates as assets of the organization, believes in the competences of the subordinates and so forth.

11.7 Conceptual Framework of Leadership, Quality of Work Life, Job Satisfaction and Performance

The conceptual framework adapted from (Subrun & Subrun, 2022a, 2022b, 2022c) undertakes that the New Facet Leadership Style is related as a successful leadership style, that can promote quality of work life, job satisfaction and performance as quality education. Leaders' leadership styles have turned out to be a key factor that governs the destiny of the institutions. The relationship in between the New Facet Leadership Style and its impact on quality of work life, job satisfaction and performance are demonstrated in Fig. 11.1. The conceptual framework that directed this research is founded on the New Facet Leadership Style and its impact on quality of work life, job satisfaction and performance. The New Facet Leadership Style as an appropriate emerging bold leadership style may create an enhanced quality of work life, job satisfaction and performance. The present conceptual framework in Fig. 11.1 is partially adapted from the conceptual framework of Subrun and Subrun (2022a, 2022b, 2022c).

11.8 Main Components Under Investigation

There is an ever-increasing need for a suitable leadership style with versatile leadership skills, which can cope with any types of challenging situations (Fig. 11.2). The leadership style in terms of the New Facet Leadership Style is adopted from Subrun and Subrun (2022a, 2022b, 2022c), responds to the needs for the appropriate leadership style. It responds to the needs of globalization and the new evolving schooling systems of today. The application of the suitable leadership style overcomes barriers (Ramesh & Subrun, 2020) in terms of New Facet leadership Style by affecting the quality of work life and satisfaction in the organizations. This leadership style is based

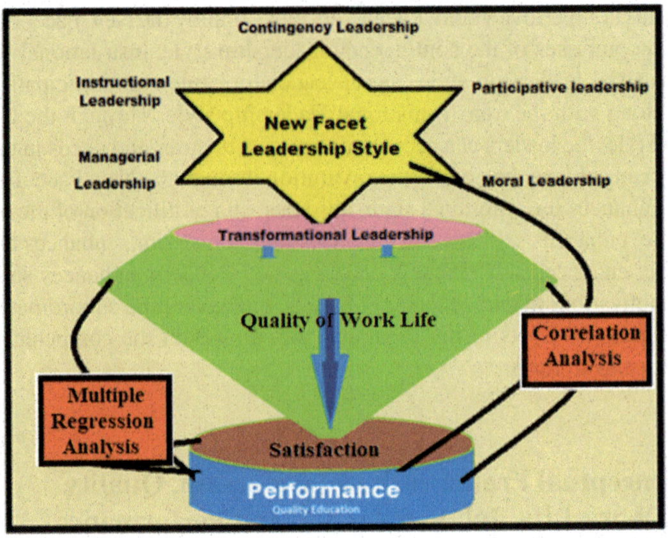

Fig. 11.1 Conceptual Framework for the Mauritian Education System on leadership, quality of work life, satisfaction and performance (*Source* Author compilation)

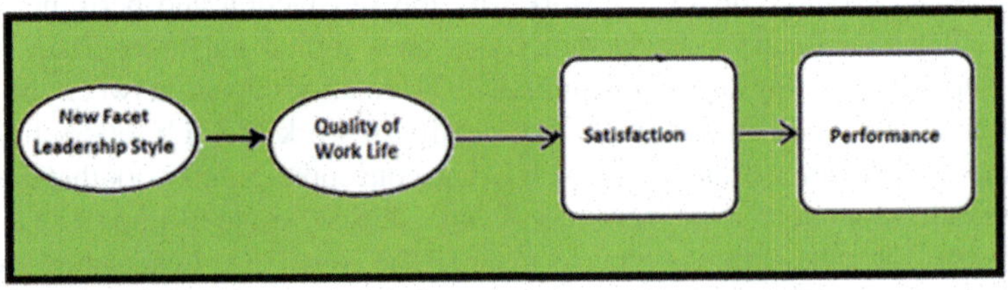

Fig. 11.2 Theoretical model (*Source* Author compilation)

on the contingency leadership theory and the transformational leadership theory comprises of contingency leadership, instructional leadership, managerial leadership, moral leadership and participative leadership along with the transformational leadership styles to handle all types of situations.

11.9 Methodology

96 rectors and 404 educators from both state and private secondary schools took part in the survey questionnaire. The statistics were scrutinized using SPSS. A sequence of analysis was carried out to investigate the impact of the New Facet Leadership Style of the secondary school leaders on quality of work life, job satisfaction and performance. The relationships were assessed through correlation analysis and multiple correlation

analysis. It analyzes the impact of the New Facet Leadership Style on the quality of work life, job satisfaction and performance in the secondary schools in Mauritius.

Null Hypothesis "There is no significant relationship between the New Facet Leadership Style, quality of work life and satisfaction in the Secondary Schools (Private and Public)".

Correlation analysis and the multiple regression analysis were conducted to test the Hypothesis and the results of the analysis are provided in Fig. 11.3. As shown in Fig. 11.3, the data reveal that there is a significant relationship between educational leadership styles (New Facet Leadership Style), quality of work life, satisfaction and performance in Secondary Schools (Private and Public). Therefore, the null Hypothesis is rejected at 1% level.

Figure 11.3 reveals that the correlation coefficient between the New Facet Leadership Style and quality of work life is 0.582 which indicates a 58.2% positive relationship between New Facet Leadership Style and quality of work life and is significant at a 1% level. Similarly, the correlation analysis between the New Facet Leadership Style and Satisfaction is 0.388 which indicates a 38.8% positive relationship between New Facet Leadership Style and satisfaction and is significant at a 1% level. 36.2% between the New Facet Leadership Style and performance, 43.1% between the quality of work life and satisfaction, 42.5% relationship between Satisfaction and Performance perception about the system, all of them are positively correlated with the New Facet Leadership Style, Quality of Work life, Satisfaction and Performance are significant at 1% level. Hence the null hypothesis is rejected at a 1% level. New Facet Leadership Style has a positive effect of 36.7% on performance, 38.8% on satisfaction, 58.2% on the quality of work life. The satisfaction leads to a substantial positive effect of 42.5% on performance. The quality of work life has an impact value of 43.1% on satisfaction. They are significant at 1% level and hence the null hypothesis is rejected at a 1% level. It is in line with the perception of Ilies and Metz (2017) study, where the latter claimed that one of the critical jobs of leaders is to persuade workforces so that they turn out to be as productive as possible. The New Facet leadership Style of the institutional leaders has a positive influence on the quality of work life which in turn has a positive effect on satisfaction and overall performance in the institutions. Subsequently, it supports the notions of Zhu, Chew & Spangler (2005) that leadership styles were connected with enhanced

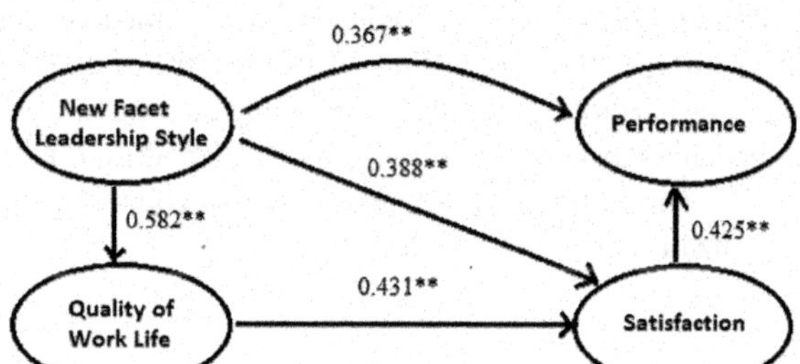

Fig. 11.3 Correlation and regression analysis for the significant relationship between New Facet Leadership Style, Quality of Work life, Satisfaction and Performance (*Source* Author compilation)

quality of work life, as leadership lies on shared actions, where the team followers work together in whatever characters of authority and supremacy they may have, not based on position supremacy. Moreover, the data reveals that satisfaction leads to a substantial positive effect of 42.5% on performance. The New Facet Leadership Style directly or indirectly positively affect the satisfaction and performance.

Thus, it may be inferred that the New Facet Leadership Style has a substantial positive effect on performance and this is in line with the notion of Ghani, Derani, Aznam, Mohamad & Zakaria (2018), that transformational leadership also contributes to intensify the motivation of staff and encourage proliferation mechanism in their attainment and performance. The New Facet Leadership Style has a positive and substantial positive effect like transformational leadership, since it encompasses a number of leadership style which enables the leaders to deal with various types of situations. Consequently, the New Facet Leadership Style encourages the subordinates to perform better. It is in line with the perception of Ilies and Metz (2017) study, where the latter claimed that one of the critical jobs of leaders is to persuade workforces so that they turn out to be as productive as possible. The New Facet leadership Style of the institutional leaders has a positive influence on the quality of work life which in turn has a positive effect on satisfaction and overall performance in the institutions. Subsequently, it supports the notions of Zhu, Chew & Spangler (2005) that leadership styles were connected with enhanced quality of work life, as leadership lies on shared actions, where the team followers work together in whatever characters of authority and supremacy they may have, not based on position supremacy. Moreover, the data reveals that satisfaction leads to a substantial positive effect of 42.5% on performance. The New Facet Leadership Style directly or indirectly positively affect the satisfaction and performance. This can be supported by the fact that satisfaction is to a great extent dependent on the type of leadership style (Lee & Lam, 2016; Tsai & Chang, 2016). The New Facet Leadership Style has a great influence on quality of work life. Quality of work life governs the various factors that affect the performance, satisfaction and other aspects of the institutions that contribute to the quality of work life (Bambale et al., 2017; Chan, 2019; Ghani et al., 2018; Lythreatis et al., 2019; Matthews et al., 2018) for teachers and other stakeholders are enhanced by the school culture under the fair supervision of a good leader. Quality of work life is produced when the work atmosphere is effective because of the contentment of the requirements of the institution and individual requirements (Secapramana & Nugroho, 2017). Thus, it may be inferred that, the New Facet Leadership Style employed by the leaders is capable to generate a conducive environment to proliferate the capabilities of the subordinates through a good quality of work life. Furthermore, with respect to leaders' behaviours, leadership style significantly affects the work results of employees, both attitudinally and behaviorally (Le & Lei, 2017). This ultimately influences job satisfaction, work performance, innovation, organizational commitment and so on.

11.10 Conclusion

The New Facet Leadership Style has a positive effect on performance (36.7%), satisfaction (38.8%) and quality of work life (58.2%). The satisfaction leads to a substantial positive effect on performance (42.5%). The quality of work life has a great impact on satisfaction (43.1%). Thus, it may be concluded that the New Facet Leadership Style has a substantial positive effect on the quality of work life, satisfaction and performance. The result is in line with the notion of Ghani, Derani, Aznam, Mohamad & Zakaria (2018), that leadership (transformational leadership) also contributes to intensify the motivation of staff and encourage proliferation mechanism in their attainment and performance. The New Facet Leadership Style has a positive and substantial positive effect like transformational leadership, since it encompasses a number of leadership style which enables the leaders to deal with various types of situations. Consequently, the New Facet Leadership Style encourages the subordinates to perform better. The New Facet Leadership Style generates a positive quality of work life for the subordinates which in turn gives rise to a conducive working environment. The conducive working environment creates satisfaction among the subordinates and thus the willingness among the subordinates to strive for a better performance occur. Thus, the data reveals and supports the facts that the New Facet Leadership Style can be the future of the institutions if the leaders in all types of institutions adopt the New Facet Leadership Style. Moreover, the New Facet leadership Style comprises a number of leadership styles, thus the leaders must employ it judiciously to counteract unforeseen situations so as to generate a better quality of work life which will lead towards job satisfaction and ultimately enhance the overall performance of the institutions to new heights.

References

Adarkwah, M. A., & Zeyuan, Y. (2020). The paradoxical relationship between principals' transformational leadership styles and teachers' motivation. *International Journal of Educational Excellence, 6*(2), 15–46.

Akar, H., & Ustuner, M. (2019). The relationships between perceptions of teachers' transformational leadership, organizational justice, organizational support and quality of work life. *International Journal of Research in Education and Science (IJRES), 5*(1), 309–322.

Alkahtani, A. H. (2016). The influence of leadership styles on organisational commitment: The moderating effect of emotional intelligence. *Business and Management Studies, 2*(1), 23–34.

Arnalod, K. A., & Loughlin, C. (2013). Integrating transformational and participative versus directive leadership theories. Examining intellectual stimulation in male and female leaders across three contexts. *Leadership & Organizational Development Journal, 34*(1), 67–84.

Atasoy, R. (2020). The relationship between school principals' leadership styles, school culture and organizational change. *International Journal of Progressive Education, 16*(5), 256–274.

Avolio, B. J. (2011). *Full range leadership development*. (2nd ed.). Sage. https://doi.org/10.4135/9781483349107

Bambale, A. J., Girei, A. A. & Barwa, T. M. (2017). Leadership styles and workers' performance in small and medium scale industries in Adamawa state, Nigeria: A research model. *Journal of Economic Development, Management, IT, Finance & Marketing, 9*(2).

Baptiste, M. (2019). No teacher left behind: The impact of principal leadership styles on teacher job satisfaction and student success. *Journal of International Education and Leadership, 9*(1), 1–11.

Barnett, D. (2018). Leadership and job satisfaction: Adjunct Faculty at a for-profit university. *International Journal of Psychology and Educational Studies, 4*(3), 53–63.

Bass, B. M. (1998). *Transformational leadership: Industrial, Military, and Educational Impact.* Lawrence Erlbaum Associates Publishers.

Bass, B. M. (2008). *The bass handbook of leadership: Theory, research & managerial application* (4th ed.). Free Press.

Bellibas, M. S., & Liu, Y. (2017). Multilevel analysis of relationship between principals' perceived practices of instructional leadership and teachers' self-efficacy perceptions. *Journal of Educational Administration, 55*(1), 49–69.

Berson, Y., & Linton, J. D. (2005). An examination of the relationships between leadership style, quality, and employee satisfaction in R&D versus administrative environments. *R&D Management, 35*, 51–60.

Bronkhorst, B., Steijn, B., & Vermeereen, B. (2013). Transformational leadership, goal setting, and work motivation: The case of a Dutch Municipality. *Review of Public Personnel Administration, 35*(2), 124–145.

Bouwmans, M., Runhaar, P., Wesselink, R., & Mulder, M. (2017). Fostering teachers' team learning: An interplay between transformational leadership and participative decision-making. *Teaching and Teacher Education, 65*, 71–80.

Boyer-Davis, S. (2018). The relationship between technology stress and leadership style: An empirical investigation. *Journal of Business and Educational Leadership, 8*(1), 48–65.

Cansoy, R. (2019). The relationship between school principals' leadership behaviours and teachers' job satisfaction: A systematic review. *International Education Studies, 12*(1), 37–52.

Chan, S. C. H. (2019). Participative leadership and job satisfaction: The mediating role of work engagement and the moderating role of fun experienced at work. *Leadership & Organization Development Journal*. https://doi.org/10.1108/LODJ-06-2018-0215

Clark, S. N., & Clark, D. C. (2000). Developmentally responsive curriculum and standards-based reform: Implication of middle level principals. *NASSP Bulletin, 84*(615), 1–13.

DeCarlo, T. E., & Agarwal, S. (1999). Influence of managerial behaviours and job autonomy on job satisfaction of industrial salespersons. *Industrial Marketing Management, 28*, 51–62.

Deluga, R. J., & Souza, J. (1991). The effects of transformational leadership style on the influencing behaviour or subordinate police officers. *Journal of Occupational Psychology, 64*, 49–55.

Demir, M. (2011). The analysis of the relationship among emotional intelligence, organizational deviance, quality of working life and turnover intentions in hospitality business. *European Journal of Tourism Research, 4*(2), 214–216.

Dipboye, R. L. (2018). Leader emergence and effectiveness in organizations. *The Emerald Review of Industrial and Organizational Psychology* (pp. 441–493). https://doi.org/10.1108/978-1-78743-785-2201081012

Donkor, A. K., & Asante, J. (2016). Instructional leadership of basic schools in Ghana: The case study of schools in Kwaebibirem District. *American International Journal of Contemporary Research, 6*(4), 65–75.

Duyan, M., & Yildiz, S. M. (2020). The effect of transformational leadership on job satisfaction: An investigation on academic staffs at faculties of sports sciences in Turkey. *European Journal of Education Studies, 7*(2), 364–373.

Fiaz, M., Su, Q., Ikram, A., & Saquib, A. (2017). Leadership styles and employees' motivation: Perspective froma an emerging economy. *Journal of Developing Areas, 51*(4), 143–156.

Fiedler, F. E. (1951). A method of objective quantification of certain countertransference attitudes. *Journal of Clinical Psychology, 7*, 101–107.

Fiedler, F. E. (1967). *A theory of leadership effectiveness*. McGraw-Hill.

Gandolfi, F., & Stone, S. (2018). Leadership, leadership styles and servant leadership. *Journal of Management Research, 18*(4), 261–269.

Ghani, F. A., Derani, N. E. S., Aznam, N., Mohamad, N., Zakaria, S. S. A., & Toolib, S. N. (2018). An empirical investigation of the relationship between transformational, transactional female leadership styles and employee engagement. *Global Business and Management Research: An International Journal., 10*(3), 724–733.

Ghaus, B., Lodhi, I., & Shakir, M. (2017). Leadership styles of academic supervisors as predictors of effectiveness, extra effort, and satisfaction: A case of Pakistan higher education. *New Horizon, 11*(2), 33–46.

Glutom, S., Baharuddin, & Fibriasari, H. (2021). Leadership styles and school performance: A study examining factors contributing to education success in Medan, Indonesia. *Eurasian Journal of Educational Research, 93*, 357–372.

Haile, A. (2017). Leadership styles at Haramaya university and implications for corporate success. *East African Journal of Sciences, 11*(1), 51–58.

Helmiatin, (2014). The effect of transformational leadership and quality of work life toward organization citizenship Behavior at Universitas Terbuka. In *11th International Annual symposium on Management*, pp. 14–17, Maret 2014, Universitas Surabaya Singhasari Resort Malang.

Heskett, J. L., Sasser, W. E., & Schlesinger, L. A. (1997). *The service profit chain*. The Free Press.

Ho, W. X., & Lin, K. W. (2015). The effect of teacher's leadership style on the outcomes of early childhood education. *International Journal of Organizational Innovation., 8*, 77–86.

Huzzard T. (2003). *The convergence of the quality of working life and competitiveness, Stockholm*: National Institute for Working Life.

Ilies, L., & Metz, D. (2017). An empirical study on the relationship between employee motivation and leadership style conducted across industrial companies in North-western region of Romania. *The Animals of the University of Oradea. Economic Sciences, Tom XXVI, 2*, 397–404.

Islam, T., Tariq, J., & Usman, B. (2018). Transformational leadership and four-dimensional commitment: Mediating role of job characteristics and moderating role of participative and directive leadership styles. *Journal of Management Development, 37*(9), 666–683.

Iverson, R., & Roy, P. (1994). A Casual model of behavioural commitment: Evidence from a study of Australian blue-collar employees. *Journal of Management, 20*(1), 15–41.

Kheirandish S. M., (2009). Relation between working life quality and performance of employees in Irankhodro Dizeli company. Master's thesis. Alame Tabatabaii University.

Kirby, G., Douglass, A., Lyskawa, J., Jones, C., & Malone, L. (2021). *Understanding Leadership in Early Care and Education: A Literature Review*. OPRE Report 2021–02. Washington, Dc: Office of Planning, Research, and Evaluation, Administration for Children and Families, U.S. Department of Health and Human Services.

Kouni, Z., Koutsoukos, M., & Panta, D. (2018). Transformational leadership and job satisfaction: The case of secondary education teachers in Greece. *Journal of Education and Training Studies, 6*(10), 158–168.

Le, P. B., & Lei, H. (2017). How far transformational leadership supports knowledge sharing: Evidence from Chinese manufacturing and service firms. *Chinese Management Studies, 11*(3), 479–497. https://doi.org/10.1108/CMS-02-2017-0039

Lee, P. T. W., & Lam, J. S. L. (2016). Developing the fifth-generation ports model in dynamic shipping and port development in the globalised economy. *Palgrave Macmillan UK, 2*, 186–210.

Lewin, K. (1945). The Research Center for Group Dynamics at Massachusetts Institute of Technology. *Sociometry, 8*, 126–135.

Li, G., Liu, H., & Luo, Y. (2018). Directive versus participative leadership: Dispositional antecedents and team consequences. *Journal of Occupational Psychology, 91*, 645–664. https://doi.org/10.1111/joop.12213

Lythreatis, S., Mostafa, A. M. S., & Wang, X. (2019). Participative identification in SMEs in the MENA region: Testing the roles of CSR perceptions and pride in membership. *Journal of Business Ethics, 156*, 635–650.

Mahzan, F. A., & Nordin, N. M. (2021). The relationship between transformational leadership practiced and the level of job satisfaction among lecturers in MARA Professional College, Malaysia. *Asian Journal of University Education, 17*(3), 154–161.

Mancha, R. M., & Yoder, C. Y. (2015). Cultural antecedents of green behavioural intent: An environmental theory of planned behaviour. *Journal of Environmental Psychology, 43*, 145–154.

Matthews, B., Daigle, J. & Houston, M. (2018). A dyadic of employee readiness and job satisfaction: does there exist a theoretical precursor to the satisfaction-performance paradigm? *International Journal of Organizational Analysis,* https://doi.org/10.1108/IJOA-01-2018-1320

Mishra, M. & Pandey, A. (2018). The impact of leadership styles on knowledge-sharing behaviour: a review of literature. *Development and Learning in Organizations: An International Journal,* https://doi.org/10.1108/DLO-06-2018-0067.

Mosadeghrad, A. M., & Larson, E. (2002). Leadership research in business and health care. *Journal of Nursing Scholarship, 34*(2), 165–171.

Northouse, P. G. (2004). *Leadership, theory and practice* (3rd ed.). Sage.

Nyberg, A., Holmberg, I., Bernin, P., Alderling, M., Akerblom, S., Widerzal-Bazyl, M., Magrin, M. E., Hasselhorn, H., Milcrarek, M., D'Angelo, G., Denk, M., Westerlund, H., & Theorell, T. (2011). Destructive managerial leadership and psychological well-being among employees in Swedish, Polish, and Italian hotels. *Work, 39*, 267–281. https://doi.org/10.3233/WOR-2011-1175

Petyer, L. J. C. (2004). The effect of organizational culture and leadership style on job satisfaction and organizational commitment. *Journal of Management Development, 23*(4), 321–338.

Ramesh, V., & Subrun, V. (2020). Educational leadership styles, secondary school performance and learner's achievement in Mauritius. *International Journal of Psychosocial Rehabilitation, 24*(4), 5006–5018.

Rifadha, M. U. F., & Sangarandeniya, Y. M. S. W. V. (2015). The impact of work life balance on job satisfaction of managerial level employees of people's bank (head office), Sri Lanka. *Journal of Management, 12*(1), 17–27.

Robbins, S. P., & Coulter, M. (2016). *Management* (14th ed.). Pearson Prentice Hall.

Schermerthorn, J. R., Hunt, J. G., Osborn, R. N., & Uhi-bien, M. (2011). *Organizational behavior.* Aisia: John Wiley & Sons (Asia) Pte Ltd.

Secapramana, V. H. & Nugroho, E. (2017, July). Quality of work life program as a humanistic perspectives on HRM. In *2017 International Conference on Organizational Innovation (ICOI 2017)* (pp. 28–33). Atlantis Press.

Seo, Y., Ko, J., & Price, J. L. (2004). The determinants of job satisfaction among hospital nurses: A model estimation in Korea. *International Journal of Nursing Studies, 41*, 437–446.

Sergiovanni, T., & Corbally, J. (1984). *Leadership and organizational culture.* University of Chicago.

Smith, P., & Peterson, M. (1988). *Leadership, organizations and culture: An event management model.* Sage.

Subrun, L., & Subrun, V. (2015a). The digital world of education in Mauritius: Adapting the Mauritian educational system with the pace of technology. *International Journal of Learning, Teaching and Educational Research, Special Issue, 13*(4), 14–19.

Subrun, V., & Subrun, L. (2015b). Provision of quality education: Mauritius in quest of quality education. *International Journal of Learning, Teaching and Educational Research, Special Issue, 13*(4), 41–48.

Subrun, V., & Subrun, L. (2022a). The New Facet Leadership Style and secondary school performance in Mauritius. *Journal of Business and Social Sciences, 4*, 1–9.

Subrun, V., & Subrun, L. (2022b). Performance a result of the impact of the New Facet Leadership Style in secondary schools of Mauritius. *International Journal of Engineering Applied Sciences and Technology, 7*(1), 67–72.

Subrun, V. & Subrun, L., (2022c). Impact of the New Facet Leadership Style on job satisfaction in the Secondary Schools of Mauritius, *Journal of Business and Social Sciences,* Vol. 7.

Susillaningsih, F. S., Komariah, M., Mediawati, A. S., & Lumbantobing, B. M. (2021). Quality of work-life among lLecturers during oline leaning in COVID-19 pandemic period: A scoping review. *Malaysian Journal of Medicine and Health Sciences, 17*, 163–166.

Toprak, M., Inandi, B., & Colak, A. L. (2015). Do leadership styles influence organizational Health? A study in educational organizations. *International Journal of Educational Methodology, 1*(1), 19–26.

Tsai, H. L., Lu, C. S., & Chang, C. C. (2016). The influence of organisational green climate on employee' green behaviours: Evidence from the eco port of Kaohsiung. *International Journal of Shipping and Transport Logistics, 9*(6), 696.

Vikaraman, S. S., Mansor, A. N., Nor, M. Y. M., Alias, B. S., & Gurusamy, V. (2021). Ethical Leadership Practices and Trust Among Public School Leaders in Malaysia.*Asian Journal of University Education (AJUE),*17(3), pp. 174 -191.

Webb, R. (2005). Leading teaching and learning in the primary school: From 'educative leadership' to 'pedagogical leadership.' *Educational Management Administration & Leadership, 33*(1), 61–74.

Wu, M. (2012). Moral leadership and work performance: Testing the mediating and interaction effects in China. *Chinese Management Studies, 6*(2), 284–299.

Yao, C., & Huang, P. (2018). Effects of leadership style on job satisfaction and intention to stay in shipping industry. *Journal of Coastal Research. Special Iss. No., 83*, 796–801.

Yu, F. T. T. (2016). Does innovation leadership enhance creativity in high-tech industries? *Leadership and Organization Development Journal*, 37(5).

Yun, S., Cox, J., & Sims, H. P. (2006). The forgotten follower: A contingency model of lidership and follower self-leadership. *Journal of Managerial Psychology, 21*(4), 374–388.

Yuvitasari, E. (2018). The improvement of in Islamic leadership towards teacher performance with competency-based and quality of work life. *International Journal of Islamic Business Ethics (IJIBE), 3*(2), 458–472.

Zeithaml, V., & Bitner, M. J. (2000). *Services marketing: Integration customer focus across the firm.* Irwin McGraw-Hill.

Zhang, Y., & Tan, C. (2021). Transformational leadership from a Neo-Daoist lens. *Philosophical Inquiry in Education, 20*(1), 26–42.

Printed by Printforce, United Kingdom